CRITICAL ISSUES
IN
POLICE TRAINING

M. R. HABERFELD, PH.D.
Associte Professor
John Jay College of Criminal Justice

Prentice
Hall

Upper Saddle River, New Jersey 07458

Library of Congress Cataloging-in-Publication Data

Haberfeld, M. R. (Maria R.)
    Critical issues in police training / M.R. Haberfeld.
        p. cm.
    Includes bibliographical references and index.
    ISBN 0-13-083709-1
      1. Police training--United States. I. Title.

HV8142.H3 2002
363.2'068'3--dc21   2001058241

Publisher: Jeff Johnston
Executive Editor: Kim Davies
Assistant Editor: Sarah Holle
Production Editor: Naomi Sysak
Production Liaison: Barbara Marttine Cappuccio
Director of Production and Manufacturing: Bruce Johnson
Managing Editor: Mary Carnis
Manufacturing Manager: Cathleen Petersen
Creative Director: Cheryl Asherman
Cover Design Coordinator: Miguel Ortiz
Cover Designer: Carey Davies
Cover Image: Courtesy of Pennsylvania Department of Corrections
Marketing Manager: Jessica Pfaff
Editorial Assistant: Korrine Dorsey
Copy Editor: Barbara Zeiders
Proofreader: Natalia Morgan
Composition: Naomi Sysak
Printing and Binding: Phoenix Book Tech

Pearson Education LTD.
Pearson Education Australia Pty, Limited
Pearson Education Singapore, Pte. Ltd.
Pearson Education North Asia Ltd.
Pearson Education Canada, Ltd.
Pearson Educación de Mexico, S.A. de C.V.
Pearson Education—Japan
Pearson Education Malaysia, Pte. Ltd.

Prentice Hall

10 9 8 7 6 5 4 3 2 1
ISBN: 0-13-083709-1

*Dedicated
to the memory of my grandmother,
Dorota Stefania Rudenska—your spirit
is always within me.*

CONTENTS

14

POLICE TRAINING IN OTHER COUNTRIES: A COMPARATIVE PERSPECTIVE   268

15

THE FUTURE OF POLICE TRAINING: SMALL, MEDIUM, OR LARGE?   298

PREFACE

> They are trained as soldiers and expect to perform as social workers . . . . well, at least 90 percent of their time, and then, in a split second—they have to be soldiers again—quite a trick!

The materials presented in this book were gathered from a host of sources: police departments around the country and around the world, police chiefs, police managers, directors of training, police officers from the highest to the lowest ranks, students, and colleagues. It has not been my intention to identify all the police departments or all the academies specifically each time I discuss a given topic. The idea behind this relative anonymity is quite simple.

While gathering the information and performing the research, I came across many people who cared deeply about the topic of law enforcement training. They were helpful, enthusiastic, open, and cooperative. They opened their offices, files, and, more important, hearts and thoughts to me—no researcher could ask for more. There is, of course, a downside to such cooperation. The idea for the book was born out of an assumption that the approach to law enforcement training is fundamentally wrong and has to be altered, enhanced, changed, or restructured in order to professionalize the idea of policing. Without a critical analysis of the findings, the "right" answers could not be provided. However, analysis frequently entails criticism—constructive criticism but nevertheless, criticism. It is not my aim to challenge or criticize any of my sources by pointing an accusatory finger at a given police department or academy. It is nevertheless my goal to present a factual picture, not just theory, in an attempt to point out the basic misconceptions. No matter how skillfully this could be approached, a number of people, police departments, and academies would be hurt in the process in a very personal way. To avoid this potentially hurtful approach, I have decided to identify the actors involved only in very rare instances and in general to refer to agencies by approximate size or state location when possible. Sometimes, however, the identity of a given training academy or department was revealed and some direct criticism offered. Nevertheless, the goal of such an approach was to provide a valuable and replicable template that can be customized to a given department or academy. In other instances all that the reader needs to know is the approximate size, type of training delivered (through an in-house, regional, or state facility) and the floating ideas—the rest is left to one's imagination and his or her passion for the idea of policing. If this passion is even remotely close to the passion I feel for this profession, there is a good chance for progress.

I would be remiss if I not mention that as I write the final pages, many of the factual descriptions in this textbook are already obsolete. Police training changes as one reads this book, but the general themes and concepts presented to the reader remain valid and important. The contribution of this book to an understanding of the nature of police work cannot be measured against the exact number of hours allocated to training or topics discussed while this research was in progress. It is highly recommended, however, that both students and instructors check, verify, and update the current training modules prior to engaging themselves in any discussion, analysis, or evaluation. Finally, if through the ideas incorporated in this book I have managed to improve the quality of life for police officers and the communities they serve, my efforts will have been well worthwhile.

Acknowledgments

Many people have helped make my dream come true. The order in which their names appear bears little, if any, correlation to the degree of their contributions. Instead, it reflects the faulty recollection of the author's memory. That being said, I have to start with my wonderful research assistant, Daniel Antonius, whose devotion, intelligence, diligence, and overall commitment to the project cannot be overstated. If it wasn't for his optimism and willingness to put other obligations aside, I would still be working on this book—thank you Daniel, you are the best writer and research assistant!

The following people contributed to the body of knowledge included in the book by providing facts, descriptions, analyses, and ideas: Richard Wolf, who contributed from his vast knowledge in the area of corrections and corrections training; Dr. Robert R. Friedmann, from Georgia State University; Chief Dennis Nowicki, from the Charlotte–Mecklenburg Police Department; Chief Richard Pendington, from the New Orleans Police Department; Chief Ed Flynn, from Arlington, Virginia; Chief David Harmon, from the Montclair Police Department; Major John O'Hare, from the Charlotte–Mecklenburg Police Department; Captain Christopher Cooch, from the Metropolitan Police Department; Lieutenant William Gundersdorf, from the Washington Township Police Department; Lieutenant Brian Onieal, from the Jersey City Police Department; Sergeant Harold Medlock, from the Charlotte-Mecklenburg Police Department; Sergeant Michael Gardner, from the Cincinnati Police Department; Patrolman Kevin Barrett from the Englewood Police Department; Patrolman Blake Munster from the New Orleans Police Department; Patrolman Gary Holsten, Jr. from South Brunswick

Township; the staff of the Southeastern Public Safety Institute at St. Petersburg Junior College; Tony Narr, from the Police Executive Research Forum; and Thomas Shaw, from the Virginia Criminal Justice Academy. A special thank you to Sergeant Peter M. Casamento III from the Maywood Police Department, who developed with me the outline for the book and shortly after had to leave the project due to personal obligations—it is *our* vision, Peter.

Special acknowledgments go to the staff of Prentice Hall, Kim Davies, whose faith in me and commitment to this project brought it to its final version; and Sarah Holle, for her never-ending willingness to help.

I also thank the reviewers of this book: Captain James F. Albrecht, from the NYPD; Professor Barry Sherman, from Madonna University; and Professor Charles E. Myers, from Aims Community College.

Finally, to my family—thank you for being supportive and understanding: my parents, Dr. Lucja Sadykiewicz and Colonel (retired) Michael Sadykiewicz; my husband, Gabriel; and my daughters, Nellie and Mia.

Many more people were involved in providing the materials and information included in this book. I did not mention their names for a number of reasons, none of which had to do with lack of appreciation—to all of you who remain nameless, my heartfelt *thank you!*

PROACTIVE TRAINING

The citizen expects police officers to have the wisdom of Solomon, the courage of David, the strength of Samson, the patience of Job, the leadership of Moses, the kindness of the Good Samaritan, the strategical training of Alexander, the faith of Daniel, the diplomacy of Lincoln, the tolerance of the Carpenter of Nazareth, and finally, an intimate knowledge of every branch of the natural, biological, and social sciences. If he had all these, he might be a good policeman! (August Vollmer, cited in Bain, 1939).

INTRODUCTION

Berkeley's legendary police chief may have been frustrated with the public's unrealistic expectations for its police officers, and his tongue was probably planted firmly in his cheek when he uttered this oft-quoted observation. One thing is quite clear, however: If August Vollmer could spend some time on the job with today's police officers, he would conclude that present-day law enforcement personnel need to be endowed with *at least* all the traits that he claimed yesteryear's citizens demanded of their officers.

Today's citizens have a romanticized notion of police work and almost no idea of what the rest of law enforcement is all about, but when the public does come in contact with law enforcement officers, they expect a lot. It is a central premise of this book that the most effective law enforcement personnel are those who are able to draw upon the characteristics on Vollmer's list.

Throughout this book, the reader will learn how creative approaches to training can assist law enforcement personnel in acquiring, developing, and enhancing skills associated with the heroic traits identified by Vollmer. Examples will be drawn from policing training, but it requires little imagination to transpose these to other law enforcement disciplines.

DEMANDS AND EXPECTATIONS: WHAT DOES THE PUBLIC WANT?

At the beginning of the twenty-first century, the public's demands on and expectations of the criminal justice system are often contradictory. People generally demand a lot but often have low expectations of law enforcement. In the 1980s, demands escalated as sharp increases in violent crime led the public to support spending more money on local policing efforts. Police chiefs responded by experimenting with varying approaches, settling (for now) on community-oriented policing as the modern approach to policing. Community-oriented policing is discussed in more detail throughout the book.

More recently, as crime rates have decreased dramatically throughout the country, public demands have shifted to exhorting the police to keep the lowered crime rates low. People demand ever-harsher penalties for convicted offenders: Truth in sentencing (legislation requiring determinate or fixed sentences for convicted offenders) and "three strikes and you're in!" (prison, that is—and for the rest of your life, with no possibility of parole) are political mantras that have found favor with the public.

NEW CHALLENGES TO EFFECTIVE LAW ENFORCEMENT

Globalization

Globalization has created new opportunities and challenges for the world's economies but also for its criminals. Technological developments have changed the way that people communicate and do business. The significance of the computer generally, and the Internet in particular, is only beginning to be understood. A new breed of technologically sophisticated criminals already present new challenges to law enforcement. As has always been the case, law enforcement visionaries need to fashion creative solutions that anticipate, rather than merely react to, ever-evolving global criminal enterprises.

Family Disintegration

The public's perceptions of the problems of social control is that it has grown more complex. People wonder whether a spate of shootings of children by children in public schools is connected to the breakup of the nuclear family, or whether violence on television or the Internet has anaesthetized young killers. The public wonders whether the proliferation of street and prison gangs means that gangs have become family substitutes.

Different "Publics," Different Perceptions

In the United States, extraordinary economic disparities between the middle class and the underclass have exacerbated tensions between the residents of poor neighborhoods in our cities and police officers, who often do not reflect the demographic makeup of the communities they serve. Different groups of people have differing perceptions of reality. One need only consider the reactions of the various "publics" when the verdict in the O. J. Simpson case was announced. Many minority-group citizens had their faith restored in the U.S. criminal justice system. They believed that an African-American defendant, who had been framed by a corrupt police conspiracy, was properly vindicated. Many white citizens, on the other hand, decried the verdict. They believed that an "obviously guilty" defendant had been wrongfully acquitted by a criminal justice system that had succumbed to the power of "the best defense that money could buy." To them, Simpson had gotten away with murder.

But the perspectives of America's largely white suburban public are quite different from those of its nonwhite urban public. The latter often view the police as an occupying force, and local residents have conflicting views about that force. They look to the police to keep them safe, but they resent many of the methods that the police employ to do so.

Diminished Expectations of Law Enforcement Officers

People expect less of individual law enforcement officers than of the police department or corrections department for which they work. People's expectations that their neighborhoods will be safe and that prisons will be secure do not mean that they expect individual officers to be heroic. Even the least cynical citizens understand that police officers are, in fact, only human.

Television reports and newspaper articles present sensational accounts of police officers shooting unarmed civilians. Allegations that police officers have taken bribes from drug dealers are assumed to be true. There seems to be a widespread belief that because officers "are only human" and must "deal with the worst elements of society," such conduct "comes with the territory" (Crank, 1998). People do not condone such behavior, but they are not shocked by it.

Yet people call upon these same officers to settle disputes with their neighbors, to roust the noisy local drunk, to keep teenagers out of the park late at night, and to help search for a missing pet. The same officer who citizens suspect may use excessive force if he encountered them in other circumstances is called upon to assist them in times of need.

PROFESSIONALIZING LAW ENFORCEMENT THROUGH TRAINING AND EDUCATION

Much can be done to better prepare law enforcement personnel for the wide array of tasks they must perform every day. Law enforcement officers can be taught self-esteem-enhancing skills that will also improve their public image. If these skills are reinforced by organizational norms adopted by our police departments, and if the departments aggressively educate their publics to begin to understand the true nature of law enforcement work and the many challenges to effective law enforcement, officers can begin to meet the demands of the publics because those demands become more realistic. Only then can the professionalizing of law enforcement begin in earnest.

What follows is a brief discussion of each of the heroic characteristics that Vollmer claimed that the public expected of police officers. Each is translated into themes that will be developed throughout the book. Each characteristic is discussed in terms of the related skills and abilities that law enforcement officers need to be able to draw upon as they do their jobs. Skills, and in many cases abilities as well, can be learned. Learning is accomplished through education and training and is reinforced by clear and consistent on-the-job reinforcement of organizational norms.

Reasonable readers will disagree with one or more of Vollmer's Biblical or historical associations. It is beyond the scope of this book to recount Bible stories or to assess history critically. Was Solomon wise or merely reckless and lucky? Is Abraham Lincoln a good example of an effective diplomat? Vollmer's choices are presented for illustrative purposes only. Some are discussed with others, as appropriate. In all cases, however, this book is concerned with the skills and abilities associated with each choice and with what can be done to teach and nurture them through law enforcement training.

Wisdom of Solomon

Life is a series of choices. For most people the choices are routine aspects of daily living that often are accomplished without any awareness that a choice is being made. Whether to get up when the alarm clock rings or to stay in bed; whether to eat toast or cereal for breakfast; whether to cross the street at the first or second intersection—each person's daily life is filled with a series of usually inconsequential choices. On-the-job decisions are usually more important. How one behaves and performs at work influences how one is perceived by one's colleagues and may determine whether one remains employed and opportunities for promotion.

But on-the-job decisions of law enforcement officers are fundamentally different. Decisions by police officers are likely to have profound implications for the people with whom they come in contact and for the officers themselves. They often affect people's liberty and personal safety. Some decisions determine whether people—citizens and officers—live or die.

Discretion in law enforcement has been a hot topic for more than 40 years, when the American Bar Foundation Reports (1956) first revealed that law enforcement involved a series of discretionary decisions. Since then, considerable research has focused on the discretionary aspects of police work (Black, 1980; Brown, 1981; Cohen, 1996; Crank, 1992, 1998; Davis, 1970; Finckenauer, 1976; H. Goldstein, 1977; J. Goldstein, 1998; Kleinig, 1996; Skolnick, 1993; Skolnick and Bayley, 1986; Sykes, 1989; Walker, 1993; Wilson, 1989; Wilson and Kelling, 1982). However, the issue can be summed up as follows: "To enforce or not to enforce, that is the question."

Laws, as well as guidelines, rules and regulations, and operating procedures developed by police departments, may vary in degree of specificity, but all share a common feature: They are inherently ambiguous. That is, each requires that a specific set of facts and circumstances be applied to the law or rule in question. Thus, no matter how precise or detailed a rule, each officer is called upon to exercise his or her discretion: first, to decide whether the facts with which the officer is confronted bring the conduct or circumstance within the ambit of the law or regulation, and then to decide whether to enforce the law or regulation. People sometimes associate the latter discretion with bad choices: the choice of some law enforcement personnel to be corrupt. The presumption is that decisions not to enforce a law or regulation are invariably the result of corruption. This is obviously not the case: Often, "discretion *is* the better part of valor."

One of the most important benefits of training is that it confers "wisdom" by instilling in officers the importance of acting on facts and by providing and enhancing the skills necessary to develop the facts. Even more fundamentally, effective training incorporates the norms and values of the organization into all curricula: Officers learn that the organization will always support correct decisions; officers learn that the organization will usually support incorrect decisions made in good faith; and officers learn that the organization will never support bad decisions made in bad faith. Thus, the use of necessary force by police or corrections officers will always be supported. The use of unnecessary force will be supported if it is applied based on a good-faith belief that it is necessary under the circumstances as the officer believes them to be. A failure to report the use of force will never be supported.

Some would say that law enforcement discretion involves the use of "common sense." But what does common sense tell a police officer about whether an intoxicated parent presents a serious-enough danger to a child that the child should be taken from the parent and taken to child welfare authorities? Training is an indispensable aid in teaching an officer how to assimilate available facts quickly so that he or she can make the best possible decision that the circumstances will allow.

Courage of David and Strength of Samson

During childhood, most people learn to avoid danger. Parents teach children not to take candy or accept rides in automobiles from strangers. People are taught not to confront danger—they are told to turn over valuables to a weapon-wielding assailant to reduce the likelihood of getting hurt. But these same people expect police officers to run toward danger (Crank, 1998). If a would-be robber is pointing a gun at a potential victim, a police officer is expected to risk his or her life to disarm the perpetrator, save the victim, and thwart the attack. Contrast these expectations with those associated with firefighters. According to the firefighter's code of conduct, in a life-threatening situation, a firefighter is expected to preserve his or her own life before that of the person the firefighter is attempting to rescue.

How do police officers generate much-needed courage? Is there some identifiable personality trait peculiar to the makeup of police and corrections officers that law enforcement recruiters can spot? What accounts for the frequent and usually unpublicized acts of courage and heroism exhibited every day by law enforcement officers?

Danger is a central theme of police work. Thinking about it and preparing for danger are critical objectives of police and corrections training. Perceptions of danger are underscored in Peace Officer Standards and Training (POST) training, where officers vicariously experience, learn, and relearn the potential for danger through "war stories" and field training after graduation from the academy (Crank, 1998). Law enforcement trainers cannot *teach* courage. Officers are not *trained* to be brave. But they are given tools which, when understood and properly utilized, enable them to perform acts of bravery that present an illusion of courage.

Trainers can help officers to develop and enhance skills that enable an officer to disarm an aggressive and violent perpetrator and to defend himself or herself in the face of danger. Physical training improves an officer's strength and endurance; weapons and tactics training develops an officer's use-of-force skills. Training and retraining keeps these skills sharp and often instills in the well-trained officer a confidence that in some circumstances may result in courageous conduct. But with confidence comes the risk that it may be misplaced, leading to foolhardy rather than courageous conduct that may lead to preventable injuries or deaths of an officer, a victim, and/or a perpetrator.

Discussions about the role of physical strength in law enforcement invariably lead to consideration of the role of female officers. Whether to use force is not a philosophical question for law enforcement officers: It is not a question of "would" or "whether" but of "when" and "how much" (Rubinstein, 1973). Indeed, a frequently heard lament of police officers is: "They teach us what not to do, but they do not teach us what to do."

Training rarely focuses on the psychological effects of uses of force on the officer. The officer's emotional response tends to be ignored or minimized, and only limited attention is paid to force's psychological impact on the recipient. Training that promotes courage must, of necessity, include curricula that provide information about the community setting in which an officer works (and about changes in that community or other setting), about the citizens with whom the officer interacts, about crime patterns—in short, any and all information that will enable an officer to best understand whatever situation he or she confronts. Training that promotes courage must also teach and reinforce information-gathering skills, so that officers can augment their information bases as they encounter specific situations, with inevitably unique circumstances.

Physical prowess alone does not equate with courage. Ignorance of the circumstances of the situation that an officer confronts may generate "blind" courage but may yield disastrous outcomes for the officer, victim, and perpetrator. Physical prowess must work in tandem with "informational prowess" so that exuberance born of confidence in one's physical prowess is tempered by the prudence born of confidence in the information that an officer possesses. It is this combination that enables officers to act bravely and with as much safety as circumstances allow. It is the responsibility of law enforcement training to promote officer courage by developing physical and informational prowess. Training that does not meet both objectives fails to meet its responsibility to the officers it develops and nurtures.

Leadership of Moses

Leadership in law enforcement encompasses two very different concepts. The first is the traditional notion of bureaucratic leadership: Moses-type leadership. Police chiefs and commissioners preside over bureaucracies that are quite similar in structure. Each "leader" needs to be able to combine strategic planning "vision" with the management skills needed to implement the vision and to lead his or her subordinates. In varying degrees, the same skills are essential leadership ingredients for high-ranking subordinates as well: deputy commissioners, precinct commanders, majors, and captains. Whether the model is total quality management (TQM), a limited participatory-management approach, or any other type of organizational decision making and planning, there seems to be a "one size fits all" approach to police leadership training.

The second, less obvious, but equally important leadership concept involves the exercise of control by law enforcement officers. Police and correction officers are vested with the authority to use coercive force, and often its use is the only means by which a situation can be controlled. But many more situations are controllable through the creative use of leadership skills: the ability to persuade.

In most walks of life, levels of education and training are society's yardsticks by which it measures one's suitability to lead and by which it distinguishes leaders from followers. Law enforcement agencies throughout the United States have focused substantial attention and resources on developing *leadership training* curricula. Whether these are provided at in-agency training academies or by outside sources, including universities, regional training centers, or national professional organizations, curricula typically target upper and middle management. Little attention has been focused on nurturing leadership skills among street-level police officers. One of the central premises of this book is that the failure to develop the leadership skills of law enforcement officers leaves many officers unprepared for managing "street" situations as they arise.

Leadership for the line officer could be defined as the ability to exercise full control in the street, or any other environment, so as to achieve a desired outcome. It is axiomatic that often the power of persuasion is more available and realistic a tool than is the power of coercive force. A police officer may sooner employ strong interpersonal skills to diffuse a potentially volatile situation on a crowded street corner than to forcibly take loud antagonists into custody—displaying a firearm may create unacceptable risks of harm to innocent bystanders. The best leadership is exhibited by the officer who is able to achieve the desired result with the least risk to the public, for whom he or she is expected, after all, to provide safety and security.

Patience of Job and Faith of Daniel

The virtues of enduring undeserved suffering with patience and with faith are central to the stories of Job and Daniel. In this respect, both stories present valuable lessons for today's law enforcement personnel. For it is no exaggeration to suggest that suffering—widely perceived (and usually correctly so) by law enforcement officers as undeserved—is an occupational reality and, for some, a daily burden.

Police officers believe that they are a "righteous" group—that they are "on the side of the angels." In their view, theirs are the only jobs requiring one to deal every day with "evil"—to see, hear, touch, and smell it (Fletcher, 1991). Above all else, law enforcement personnel are expected to endure without complaining the consequences of daily exposure to society's most vexing problems. Indeed, they know that citizens expect them to "handle it" because society pays them to do just that. But until recently, law enforcement officers were expected to handle it alone. Belatedly, law enforcement agencies began to appreciate the need to develop strategies to enable officers to cope with job stress. In our view, most approaches are reactive—too little, too late.

A pioneering stress management program was developed in 1985 by the Royal Canadian Mounted Police when its chief psychologist recommended the creation of a center to provide psychological services to recruits and veteran officers. Until then, most law enforcement agencies had attempted to deal with individual cases of apparent job stress as it was "diagnosed" (Thibault et al., 2001).

Over the past 10 years, *employee assistance programs* (EAPs) have emerged in police departments around the country as a now-obvious and well-intentioned management response to job stress. Typically, EAPs are staffed by peer officers who have received training in basic counseling and in referral procedures. Larger departments sometimes employ mental health professionals, including psychiatrists. In many ways, EAPs are analogous to community walk-in mental health clinics, where citizens who exhibit mental health symptoms can receive professional counseling and care.

Law enforcement personnel need more. They have to deal with dead bodies—from shooting victims on the street, to drowning victim "floaters," to suicides, to victims of child abuse and other atrocities. Law enforcement officers often are called upon to notify and to comfort next-of-kin, but no one comforts them. They know that the next encounter they have with a citizen or a criminal may be fraught with danger. Law enforcement officers are subjected to stress with a frequency and to a degree that most citizens would not and could not tolerate. It is unsurprising, then, that many officers resign in favor of less stressful jobs. Too many officers have neither the patience nor the faith to endure the seemingly endless exposure to stress.

Law enforcement agencies must proactively protect their human resource assets. High employee turnover can be reduced if agencies attack its causes. Realistic assessments of on-the-job stress-provoking incidents should be incorporated into basic training academy curricula. Coping mechanisms need to be emphasized early in training. Recruits must learn to recognize the signs and symptoms of stress. They must be taught that some manifestations of the consequences of job-related stress are the norm, not the exception, and that there is nothing shameful about seeking assistance in coping with stress-related problems. These notions can be taught in basic academy training and should be reinforced regularly in in-service training.

Kindness of the Good Samaritan

If one were to ask 1000 randomly selected citizens to list attributes they expected law enforcement officers to possess, it is unlikely that "kindness" would be mentioned by many, if by anyone. Paramilitary organizations such as police departments are expected to attract applicants and then to produce officers who are "professional." Kindness is something of a stretch.

Three words painted on the doors of New York City's police patrol cars exemplify NYPD's goal for the demeanor of its officers and their dealings with New Yorkers: *courtesy, professionalism*, and *respect* (CPR). Indeed, one hardly requires a public-relations CPR campaign to understand that police officers are expected to exhibit courtesy and respect toward citizens.

As the reader might expect, this is an important theme that is discussed throughout the book, as it is particularly susceptible to change through training, and is emphasized in Chapter 2. In that chapter we depict the nature of police work historically, as far removed from the CPR concept. For now, however, let us turn briefly to the difficult role of the police officer in the community.

Citizens expect courtesy and respect from their police officers. Even in the face of adversity, citizens expect police officers to display good spirits and self-control. No matter how insulting, provocative, or intimidating members of the public may be, police officers are supposed to remain "calm, cool, and collected." This is the professionalism component of CPR. Paradoxically, professionalism in the face of adversity is *demanded* of police officers, even though many officers receive fewer hours of training than that required for certification as a beautician. Despite minimal preparation, police officers are expected to be courteous and respectful to everyone—from small children and old ladies to accused serial killers.

There is no reciprocity of "kindness" or professionalism. A deranged man may insult the race or family of a police officer; he may throw bodily waste at the officer or assault the officer with a pipe. The police officer is, nonetheless, expected to suppress anger, to hold in check instinctive responses, and to display no emotions. But the expectation is unrealistic when unaided by methods of maintaining self-control when other citizens are not held to as high a standard. Police officers are only human—they need help to master self-control in the face of adversity. Appropriate training can go a long way toward providing officers with hard-to-acquire coping skills. The Good Samaritans were ridiculed as naive. Proper training can enable officers to be courteous and respectful even as they generate courtesy and respect toward the profession they represent.

Strategical Training of Alexander

Alexander the Great conjures up the image of the ultimate warrior. To the extent that movies and television glorify the "warrior" aspects of law enforcement, they reinforce an outdated notion which presumes that paramilitary structure and, more fundamentally, regimented militaristic approaches to problem solving represent the ideal.

The "disconnect" between the public's idealized notion of the warrior–police officer in a paramilitary police department is belied by the philosophy of community-oriented policing (COP) and related training concepts. Overblown, largely political descriptive rhetoric (e.g., the "war on crime," the "war on drugs") reinforce the police officer's self-image of the at-home soldier. The realities of three recurring themes—suspicion, unpredictability, and danger—cause police officers to conclude that everyone "out there" is a potential enemy (Crank, 1998).

The analogy of the law enforcement officer to the soldier is buttressed by the fact that law enforcement personnel suffer from "battle fatigue syndrome" much as soldiers do (Haberfeld, 1998). Whether this evidences itself in cynicism toward the goals of the organization to which the officer belongs, or the criminal justice system, or even the public's often-contradictory demands and expectations of its law enforcement personnel, unchecked cynicism evolves into what will be referred to as *moral corruption*. The emergence of stress management training and secondarily, employee assistance programs are the only tangible indications that some law enforcement agencies recognize and are attempting to address the serious problems presented by moral corruption.

Diplomacy of Lincoln

Every day on the job, law enforcement officers confront situations that would challenge the negotiating abilities of even the most highly trained and experienced diplomats. For police officers, diplomatic skills equate with the power of persuasion. Here we are not talking about "power" in the conventional physical sense. Although a police officer has a gun, drawing a firearm often means that the officer has been successful in the use of communication skills to defuse a situation (except, of course, when the officer is confronted with deadly physical force). Effective communication skills are critical and indispensable tools for police officers.

Bittner (1970) observed that the public calls upon the police to intercede in situations when "something ought not to be happening about which something should be done NOW." Often, the situation is not an illegal situation per se. It may involve disputes among neighbors: for example, order maintenance issues such as when one citizen objects to the volume of a neighbor's radio or to the constant barking of a neighbor's dog. Other situations may test the officer's communication skills even further, particularly if the officer is confronted with a situation that produces ambivalent feelings in the officer. For example, a police officer may sympathize with a crowd protesting the views of the legally congregating members of a hate group such as the Ku Klux Klan. If the protesters interfere with the KKK's freedom of assembly—ensured by a permit obtained from the government—the officer is obliged to take actions to prevent the protesters from interfering unduly with the KKK's lawful demonstration. Safeguarding hateful people and enforcing unpopular laws requires police officers to be able to persuade citizens to refrain from certain actions or to behave in ways about which the officer himself or herself may have doubts. But the importance of communication skills in law enforcement is by no means limited to ambivalent politically charged situations.

It is relatively easy to teach law enforcement officers the details of the laws they are obligated to uphold. The more difficult challenge to trainers is to develop curricula that produce law enforcement personnel who are trained in the diplomatic arts of reasoning and persuasion. Only when officers possess effective communication skills can they substitute the power of persuasion for physical power and thereby reduce the risk of physical injury to themselves and the people with whom they must deal.

Tolerance of the Carpenter of Nazareth

Diversity among citizenry in the United States is understood to extend beyond racial, ethnic, and gender differences to include sexual orientation and diverse political viewpoints. More than just about any other group in society, law enforcement personnel are supposed to epitomize tolerance. Law enforcement officers interact with all citizens, so tolerance means more than just "putting up" with different traits, characteristics, behaviors, and views. It includes empathy and multi-cultural orientation, referred to by Shusta et al. (1995) as *multicultural law enforcement*.

Understanding the customs and mores of the citizen groups with which a police officer comes in contact can go a long way toward enabling that officer to interpret properly information and actions critical to sound, often on-the-spot decision making. Whether street activity is viewed as suspicious or threatening to an officer may well turn on who is engaged in that activity and what the observing officers knows, or thinks that he or she knows, about the actor.

Beginning in the mid-1980s, diversity training was incorporated into most law enforcement training curricula. However, the timing of diversity training and the subjects emphasized vary dramatically from jurisdiction to jurisdiction. Agencies serving rural, predominantly white populations are unlikely to include large training blocks teaching recruits to understand and empathize with minority-group citizens who do not live in the geographical area of the agency. In contrast, large metropolitan police and correction departments are likely to include training stressing the uniqueness of the various resident populations found in many U.S. cities. Unfortunately, most agencies do not introduce diversity training until near the end of academy training. Some police departments do not provide diversity training until "rookies" are already in uniform, armed at least physically and presumably "ready to go." By waiting until the end of academy training to intro-duce diversity training, agencies create the impression that diversity training is not terribly important. If this impression is internalized by officers, the valuable lessons of diversity training may be lost.

Diversity training has become increasingly important as the faces of law enforcement have changed. No longer is law enforcement the sole province of the white male; women and minorities are recognizing the attractiveness of careers in law enforcement. As agencies become less homogeneous, officers must learn more than tolerance: Professionalism means treating one's co-workers with dignity and respect regardless of their race, ethnic background, gender, or other distinguishing characteristics.

On the street, in the precinct house, or in prisons and jails, multicultural law enforcement is about more than just being tolerant of visible and obvious differences among people. Ideally, tolerance must extend to the infinite range of values and norms, behaviors, and personal habits that differ from those with which an officer is comfortable or accustomed to. The perception that intolerance has caused an officer to abuse his or her office by violating the rights of someone who looks or thinks differently from the officer can divide a community and seriously erode public confidence in the law enforcement agency. Teaching officers how to enable one's views and beliefs to coexist with the different views and beliefs of others citizens—and other officers—is one of the greatest challenges to law enforcement training.

Intimate Knowledge of Every Branch of the Sciences

Law enforcement officers are the new generalists. Unlike professionals whose training sharply defines the parameters of their expertise, law enforcement officers have seemingly infinitely elastic job descriptions. Indeed, doctors possess ever-narrowing areas of expertise. An orthopedist would not attempt to treat—or even try to diagnose—a patient's liver disorder; an ophthalmologist is not expected to treat high blood pressure. A trusts and estates lawyer does take medical malpractice cases to trial. Only admiralty lawyers handle admiralty cases.

The "Marcus Welby" family doctor and the all-purpose attorney-generalist are dying breeds. As many professions have evolved into subgroups of highly trained specialists, law enforcement officers have evolved into ever-broadened generalists, who must instantly answer a wide range of difficult questions and take prompt and correct action:

- How should one evacuate people from a building flattened by a terrorist's bomb?
- Is the person before me suffering from cocaine shock, or is he a common drunk?
- Does the prisoner who says he's going to kill himself mean it?
- How does one restrain an autistic child?
- How does one talk to a juvenile who may have killed his parents?
- How does one control gang activity on the streets?

- How does one effectively use laptop computers, video cameras, recording devices, and cell phones to better patrol, control, and investigate?
- How do we "protect and serve," and when necessary, arrest and detain citizens of an ever-more-diverse multicultural society?

This is not meant to be an all-inclusive list. Each reader and each police officer could think of many, many others. The fact is that the public expects its police to handle almost any problem that arises.

Ever-rising expectations that law enforcement keep abreast of technological developments, cultural idiosyncrasies, and medical and mental health diagnoses—to name but a few enormous areas of knowledge—have led some of today's law enforcement strategic planners to devise unrealistic and potentially dangerous shortcuts as substitutes for knowledge acquired through training and experience. "Problem-solving" software, together with a laptop computer, are being touted in some quarters as the answer, or at least a partial answer, to the ever-escalating demands on law enforcement that officers possess the substantive knowledge to answer any question, including all those on our list. The software requires only that the officer punch in answers to a series of questions: The software will enable the computer to identify the problem for the officer and to tell the officer how to solve the problem. The prospect of such a tool, used by undereducated officers to make potentially life-and-death decisions, is too frightening a notion to contemplate.

SUMMARY

There is no quick-and-easy substitute for education and training. Who teaches and trains, and decisions about curricula and resource allocation, will shape the future of law enforcement and determine the extent to which officers will be able to respond to an increasing array of complicated and sophisticated problems. Education, academy training, in-service training, and continuing education—all are essential to ensuring that the new generalist law enforcement officers are equipped with the information and skills to meet the demands of the new millennium. The most important theme of this book, however, is the concept of proactive training. Much has been said about the need for proactive policing, a term used frequently within the "lexicon" of community-oriented policing—*proactive* as being preventive, attempting to get at the root of the problem before it develops and ultimately causes damage. This preventive approach to policing, although extremely noble in concept, has limited chances for success and long life if the grounds for implementation are not laid by utilizing the approach of proactive training. As this book begins with a historical overview of the role of police forces in our society, it sets the scene for the dire need for a change—and this change can be fully accomplished only by rethinking, in the most fundamental manner, the ways in which we offer police training in the twenty-first century.

REFERENCES

American Bar Foundation Reports (1956). In J. Kleining (ed.) (1996), *Handled with Discretion: Ethical Issues in Police Decision Making*. Lanham, MD: Rowman & Littlefield.

Bain, R. (1939). The policeman on the beat. *Science Monthly, 48*, 5.

Bittner, E. (1970). *The Function of the Police in Modern Society*. Washington, DC: U.S. Government Printing Office.

Black, D. (1980). *The Manners and Customs of the Police*. San Diego, CA: Academic Press.

Brown, M. (1981). *Working the Street: Police Discretion and the Dilemmas of Reform*. New York: Russell Sage Foundation.

Cohen, H. (1996). Police discretion and police objectivity. In J. Kleining (ed.), *Handled with Discretion: Ethical Issues in Police Decision Making*. Lanham, MD: Rowman & Littlefield.

Crank, J. P. (1992). Police style and legally serious crime: a contextual analysis of eight municipal police organizations in Illinois. *Journal of Criminal Justice, 19*, 339–350.

Crank, J. (1998). *Understanding Police Culture*. Cincinnati, OH: Anderson Publishing.

Davis, K. C. (1970). *Discretionary Justice: A Preliminary Inquiry*. Baton Rouge, LA: Louisiana State University Press.

Finckenauer, J. (1976). Higher education and police discretion. *Journal of Police Science and Administration, 3*, 450–457.

Fletcher, C. (1991). *Pure Cop*. New York: St. Martin's Paperbacks.

Goldstein, H. (1977). *Policing a Free Society*. Cambridge, MA: Ballinger.

Goldstein, J. (1998). Police discretion not to invoke the criminal process: low visibility decisions in the administration of justice. In G. Cole and M. Gertz (eds.), *The Criminal Justice System: Politics and Policies*, 7th ed. Belmont, CA: West/Wadsworth, pp. 85–102.

Haberfeld, M. R. (1998). Confronting the challenges of professionalism in law enforcement through implementation of the field training officer program in three American police departments. *Turkish Journal of Police Studies, 1*(2).

Kleinig, J. (1996). Handling discretion with discretion. In J. Kleining (ed.), *Handled with Discretion: Ethical Issues in Police Decision Making*. Lanham, MD: Rowman & Littlefield.

Rubinstein, J. (1973). *City Police*. New York: Farrar, Straus, & Giroux.

Shusta, R. M., Levine, D. R., Harris, P. R., and Wong, H. Z. (1995). *Multicultural Law Enforcement: Strategies for Peacekeeping in a Diverse Society*. Upper Saddle River, NJ: Prentice Hall.

Skolnick, J. (1993). *Justice without a Trial*, 3rd ed. Upper Saddle River, NJ: Prentice Hall.

Skolnick, J., and Bayley, D. (1986). *The New Blue Line: Police Innovation in Six American Cities*. New York: Free Press.

Sykes, G. (1989). The functional nature of police reform: the "myth" of controlling the police. In R. Dunham and G. Alpert (eds.), *Critical Issues in Policing*. Prospect Heights, IL: Waveland Press, pp. 286–297.

Thibault, A. T., Lynch, L. M., and McBride, R. B. (2001). *Proactive Police Management*, 5th ed. Upper Saddle River, NJ: Prentice Hall.

Walker, S. (1993). *Taming the System: The Control of Discretion in Criminal Justice, 1950–1990*. New York: Oxford University Press.

Wilson, J. Q. (1989). *Bureaucracy: What Government Agencies Do and Why They Do It*. New York: Basic Books.

Wilson, J. Q., and Kelling, G. (1982). The police and neighborhood safety: broken windows. *Atlantic Monthly, 249*, March, pp. 29–38.

HISTORY OF POLICING

INTRODUCTION

This chapter differs somewhat from any other chapter on the history of policing that a reader would find in a textbook. The purpose of this review is to highlight one very important theme, which is the basis for an understanding of the critical issues in police training. This theme is a premise that police forces, throughout history, served and protected the ruler, the king, a politician, but never the public. The safety and security of the public was always secondary to the safety and security of the ruler, king, politician, and so on. One cannot lead a prosperous life—as a ruler, leader, king, or politician—if the status quo is disrupted in one way or another. A happy public is a quiet public. The Roman emperors knew how to keep the public compliance—"give them bread and games, that will make them happy"—and just in case it will not, establish a strong police force.

If by looking at this brief overview, one can establish that this was indeed the case, a much more profound understanding of how and why police forces were trained will follow and will be culminated with an understanding of how police officers who are supposed to take upon a new role—one of serving and protecting the public—should be trained.

Origin of the Word "Police."   The word *police* stems from the Greek word *polis*, which refers to the citadel or government center of the city-state in Ancient Greece. In modern times *police* was first defined as any kind of planning for improving or ordering communal existence. Later it became associated with the regulation and governing of a city or a country, in regard to its inhabitants. Today, the word *police* has an exclusive meaning throughout the world.

REITH'S FOUR PHASES OF EVOLUTION

Since the beginning of recorded time, people have sought protection from danger. Reith (1975) outlines four phases in the evolution that, in its final stage, brought about establishment of the first police force.

- *Phase I*. People came together to form small communities, predominantly to ease the food findings but also to achieve a greater sense of security.
- *Phase II*. The need for laws is discovered.
- *Phase III*. The "rule breakers" emerge.
- *Phase IV*. In one form or another, means to compel the observance of laws were established.

According to Reith, more communities have perished because they could not enforce their laws than have been destroyed by natural disasters or hostile aggression. His main premise is that past civilizations fell because no police mechanism existed between the army of the ruler and the people. The country then fell into anarchy, armed troops were called to restore order, but they could secure only temporary relief, which eventually led to another outburst.

Since not much is known about the exact role and function of this first police force that Reith talks about, an interpretation of his premise is in order. It can safely be assumed that when people came together to form a community to facilitate food gathering and safety and security, from a very early stage there was some division of work based on individual skills. Furthermore, it can be assumed that there was either one or a group of leaders in a community who were "first among equals." The day on which the people discovered the first rule breaker, who might have awakened one morning and decided that as of that day he or she did not intend to take part in gathering the berries, the means to compel that person to cooperate were discovered as well. Bittner's (1970) "something ought not to be happening about which something ought to be done NOW" presents our ancestors with a dilemma: whether to employ psychological or physical means. Even if one is optimistic enough to presume that at that early stage of human development, psychological skills were available, it can safely be assumed that physical means were the ones that prevailed.

Now it is again an assumption that not every member of the community was involved in the process of convincing the "berry gatherer" that it was time to go to work. The leader or a group of leaders were probably involved. And at this point the interpretation of what really happened becomes a fertile ground for speculation. However, one may speculate that the first leaders of the prehistoric tribes must have been more intelligent (and probably stronger) than others. Even if their leadership skills rested primarily on their physical strength, it is rather doubtful that they would have liked to be involved in daily rituals of convincing the rule breakers to comply. It would be simply too overbearing and just plain dangerous. The time had come to designate somebody to do it for them (thousands of years later we meet the "village idiot" in medieval England, who would perform this task for the more mentally able). The job was certainly not very appealing, but chances are that it required physical strength. How does one make this job offer appealing? One way is by delegating some of one's authority to the designated "volunteer." In essence, when the chosen employee took upon himself or herself this first law enforcement assignment, first and foremost he or she took on the responsibility of protecting the leader—from failure to enforce and convince the rebellious "berry gatherer" to go back to work and from the potential physical danger that the rebel could have inflicted on the leader. Enforcement of the law— gathering the berries, for example—was, of course, beneficial for the entire community, but was first and foremost beneficial for the leader. The community was thus engaged in a productive activity that not only brought in food for members but for the leader as well, who might have been less intensely involved in labor exhaustive activities.

The foregoing are, of course, speculations only and not established facts. However, readers are invited to form their own opinions about the development of the first police forces and the role of police forces throughout the centuries, bearing in mind these aforementioned themes.

THE FIRST RECORDED EVIDENCE OF POLICING

The police system as we know it today has its origin in early tribal history. It is known that some tribal chiefs would appoint certain persons to assist them in both intimate and administrative duties. The reason was that these chiefs did not trust their underchiefs or the tribe as a whole. The appointees functioned as effective policemen by solving problems quickly and inflicting law observance in the clan.

It appears that tribal invaders who founded the kingdom of Sumer, and later founded Babylon, developed the first system of encoded laws. These laws were kept preserved when the Semites conquered the kingdom. Hammurabi, king of Babylonia in 2181–2123 B.C., built his laws on those of the Semites. Hammurabi also instituted a well-organized judicial system. These findings serve as evidence for the early existence of a police system, since some sort of structure had to be established to enforce the laws that were made.

Hammurabi's police force is described in the Biblical book of Genesis, in the story of the battle of four kings against five, and scholarly churchmen agree on the validity of the police force. Reith (1975) describes this police force as a "very good one . . . equal to many systems found in the country districts of Britain today." Hammurabi's police can be seen as the first ruler-appointed police in an institutional form.

Tribal history in Egypt is believed to stem from Asia via the kingdom of Sumer. There is some evidence of policing, but almost nothing is known. In the fourth dynasty there is evidence of officials with the title *tep-kher-neset* (which means "first under the king"), who were directly responsible to the king and were a part of the governmental system. The efficiency of this government suggests that a police force must have been present to keep order. In the fifth dynasty, inability to enforce the laws resulted in internal weakness and anarchy. Law and order was restored by the founder of the twelfth dynasty, Amenemhet III. However, this dynasty eventually collapsed as well. In 1580 B.C., King Ahmose established the Theban dynasty. He brought in a well-organized army and created an efficient and centralized government. Ahmose also instituted the "citizens of the army," which functioned as the natural support of the Pharaoh. This army was essentially a police force and was necessary for successful functioning of the administration and for effective internal law and order. The Theban dynasty eventually collapsed because of military neglect.

Around this time, history books also mention a police patrol in the southern Nile region. There is also evidence of police in the Assyrian and Persian empires, but it is extremely limited. In the Assyrian empire there appear to have been laws that were enforced effectively. The king was the headstone, but under him there were men called the "king's officers," and these are believed to have resembled a police force. In the history of the Persian Empire there is one indication of an institutional police system: the function of a remarkable road and postal system.

POLICING IN ANCIENT GREECE

In the early history of the Greeks, tribal organization flourished. A form of kinship police was developed which functioned so effectively together with tribal custom and traditions that laws were not necessary. Damages done to a person and/or another's property were avenged by the person or the person's kin, according to the form of punishment tolerated. The laws would develop later, when more people settled and civilization developed.

When the city-state of Athens was established, a mass of laws was instituted. In addition, there were magistrates, and courts with juries consisting of up to thousands of citizens. There were jailers and executioners who carried out sentences, but there were no institutional police forces. The kin-police system evolved. "The people were the police; the police were the people."

This changed with the dictator Peisistratus, who took over control of Athens after internal strife had resulted in anarchy. He was a person who believed in a law enforcement regime and understood that a military force would be unavailable to emergency conflicts within the city and therefore created the first institutional police in Athens, naming them the "Scythian foot-archers."

This police force consisted of slaves owned by the community (often, they would later be freed). Their weapon, a bow, seems to have been purely symbolic. The free Athenians considered police duty so degrading that they would rather be arrested by a slave than do such despicable work themselves. The police lived in tents and served as guards in the city and as assistants and ushers in the assembly and other public places. In addition, they functioned as road police, performing duties along the highways.

The city-state Sparta, an early example of a totalitarian state, had a ruler-appointed institutional police system. This police force has been called the "secret police system of Sparta" by modern historians because of the way it functioned. The Spartan police consisted of young men under the command of an *ephor*. Their main duty was to trace and kidnap slaves who were in opposition to the rulers and, if necessary, to execute them. There is evidence of massacres of immense numbers of slaves committed by this secret Spartan police force.

All the city-states in Ancient Greece (more than 150 states) succeeded in instituting some sort of police system. When Alexander the Great conquered the world, their institutional police system spread to Egypt with the Greek soldiers. The soldiers had been placed in Egypt to protect against future rebellions and were therefore given some land. The soldiers soon became the dominant element, and many Greek settlers came to establish new ground. They brought with them the institutional police system, which was effectively instituted by the king of Egypt. Eventually, it would be this law enforcement system on which the Roman emperor, Augustus, would base his empire (Engels, 1942).

POLICING IN ANCIENT ROME

The Roman Emperor Augustus Caesar was not the first to introduce a policing force to the world. Since the beginning of the development of structured cities, certain rulers have had special forces or separate branches of their armies that specialized in keeping law and order in the cities. However, Augustus was the first ruler who made a distinct police force, with different clothing, living style, and commanders, separate from the army. When Augustus became the undisputed ruler of the Roman world in 31 B.C., he was confronted with a lack of law and order. He removed many of the problems successfully, but he also found it necessary to form a body of men who could enforce the peace in Rome. An army would be too frightening to the public, so, instead, he established the Praetorian Guard in 27 B.C. (Echols, 1957). This force consisted of nine cohorts, each with 500 men, controlled directly by Augustus. *Their prime function was to protect the emperor* and enforce law and order in Rome and Italy. To distinguish the Praetorian Guard from the military, the men were quartered in private and lodging houses instead of barracks. Also, the policing guard wore togas, not uniforms, with their swords hidden under their clothing (Davies, 1977).

Because those men in the Praetorian Guard were recruited from the army, they were still frightening to the public. Therefore, Augustus in A.D. 6 introduced the Vigiles (Reynolds, 1926). This was a night watch force which both kept law and order and served as firefighters. The Vigiles were recruited from freedmen, a category that was excluded from military service, and they would live exclusively in their private homes. The Vigiles consisted of seven cohorts, each of 1000 men. Both the Praetorian Guard and the Vigiles were used as a police force throughout Italy. However, in A.D. 13, Augustus established a force specifically for law enforcement in Rome, the Urban Cohorts (Echols, 1961). This force was built up as three cohorts, each consisting of 500 men, and the early task was to police the capital. Each cohort was commanded by a captain and contained six companies. The men were all recruited from citizens. In the beginning, Augustus was in charge of the Urban Cohorts; however, he wanted somebody who could be with the force permanently, and as Calpurnius Piso became the first permanent city prefect in A.D. 13. Piso held the office as the first police commissioner until his death 20 years later (Davies, 1977).

Overall, the main function of this first police force was to supervise markets and assemblies, help local communities, guard places, arrest robbers, and escort convoys. However, when Augustus died in A.D. 14, subsequent emperors continued and expanded the policing system. Augustus's police force never became a law enforcement agency in the modern sense, but it should receive praise. This policing force was a predecessor to other police establishments that lasted several centuries with only minor changes. The relatively small force established reasonable law and order in a city which consisted approximately of 1 million people.

POLICING IN THE ISLAMIC EMPIRE

The Islamic community was founded in the seventh century A.D. by Mohammed, an Arab. Economic difficulties had led Mohammed and his followers to organize brigandage against the caravans of his enemies at Mecca. His constant success in battle gave him power and a fanatical devotion from those who fought for him. Mohammed fought and beat one tribe after the other and instituted simple laws which were easy to follow, and as a result, a form of kin-police developed. Eventually, he returned triumphantly to Mecca, where he began a plan for a holy war.

The holy war was conducted later, after the death of Mohammed in A.D. 640, by the caliph Omar. He conquered Palestine, Syria, Mesopotamia, Assyria, Babylonia, and Egypt and kept the old kin-police. However, this system could not keep law and order, and under the caliph Abu'l-Abbas, an institutional, ruler-appointed police force was established. The caliph appointed a minister, the vizir, to control the police system. Under him were a chief judge and a chief of police. The duties of the police consisted of upholding civil and religious law, being a local sensor of morals, preventing fraud, supervising commercial transactions, inspecting the quality of goods, preventing the sale of wine, preventing the playing of musical instruments in public places, and protecting slaves when they were entitled to protection. The police also had to ensure the carrying out of ritual observances, punishing people for breaking the rules of the feast of Ramadan, and making sure that people kept no items that resembled heathen idols. If anyone were to break any of the laws, it would be the duty of the police to bring the person before a judge for trial.

The Islamic empire lasted little more than a century.

POLICING IN TRIBAL AFRICA

Evidence from archaeological findings in Africa points to numerous cases of rather sophisticated societies in which an effective police system must have been a necessity. Mair has described several East African societies in which kinship security existed (Mair, 1964). For example, for the Hutu tribe in Ruanda, it was especially important to provide protection against marauders and to secure the cattle. Another tribe, the Getutu from Kenya, took in refugees from related tribes, who in return for food and shelter would serve as soldiers and guards.

In the well-documented Zulu tribe, policing had a high priority. Originally, there were two types of security societies. One acted as a check on the power of the chief, and the other acted on behalf of the chief. However, under the leadership of Dingiswayo, all security came in under the chief. This force would carry out police functions, such as fighting battles for the chief, confiscating property in the chief's name, cultivating the fields of the chief, and maintaining his kraals. In the eighteenth-century Buganda Kingdom, located at the northern shore of Uganda, the king had bodyguards, his own secret police, and executioners (Sacks, 1979).

EARLY ENGLISH POLICE HISTORY

Anglo-Saxon influence in England began with numerous raids in the fourth century. By A.D. 410, the Romans had been driven out of England and the Anglo-Saxons had taken over certain regions of the country (the Celts and Norsemen had also occupied some regions). Eventually, it would be this group that would give English culture its definitive character. The Anglo-Saxons brought with them a form of government that can be characterized as autocratic kingship. Every tribe had a king who was regarded as descended from the gods, and under him there were many grades of rank.

By the sixth century the Anglo-Saxons had established themselves in the western half of England, and within this community were townships regulated by lords. The lords stood under the church, to which they had to pay taxes. However, even though the lords functioned as regulators for the community, security was still based on kinship principles. The tribe would provide security for each other, and if a law were broken, the guilty person(s) would be punished by the tribe's rules.

In the eighth century, kinship-based policing changed. The English kings had converted to Christianity in the seventh century, and as a result, the church became more involved in politics. This resulted in a change in policing. The primary responsibility of security in the Anglo-Saxon communities was placed on the lords. The English government was thus in place, controlled by three powers: the king, the bishops who composed the king's laws, and the lords who administered the laws. The Anglo-Saxons brought with them to England acceptance of mutual responsibility for civil and military protection of individuals. Groups of 10 families, called *tithings*, banned together to provide collective responsibility for maintaining local law and order.

When the Normans conquered England in 1066, they accepted the Anglo-Saxon legal system. This system was expanded into the *frankpledge system*, under which the king demanded that all free Englishmen swear to maintain peace. Furthermore, the system was centralized under King William (reigned

1066–1087), who divided the land into shires. Frankpledge was a police system developed by the conquering Norman monarchy as an instrument of central government control. Prior to the conquest, policing and the administration of justice were almost totally under the control of local lords and nobles. Frankpledge sought to take that control out of local hands and place it in the hands of the king (Klockars, 1985).

The frankpledge system was replaced by the *parish constable system.* The sheriff became the most powerful royal officer of the shire and also supervised the court. Under him he had constables, whose main work was the making of presentments to the courts. The origin of the words *sheriff* and *constable* dates back to about A.D. 1200. At this time law enforcement was poorly organized and resembled more closely an organized posse. In 1285 the Statute of Winchester was enacted, which established three practical measures: The first was *watch and ward*, which provided town watchmen to patrol a city during the night, to supplement the traditional daytime duties of the constables. Next, when a person committed a crime, and after he or she was identified, a *hue and cry* would be raised and a group would be organized to track down the offender. The group would be lead by the "shire reeve," the "leader of the county," or by his mounted officers, the "comes stabuli" (Schmalleger, 1993). (It is from these two words that our modern words *sheriff* and *constable* are derived.) The third measure, *assize of arms*, required every male between the ages of 15 and 60 to keep a weapon in his home.

The Statute of Winchester provided that one man from each parish would serve a one-year term as a parish constable on a rotating basis. Each parish constable was responsible for organizing the group of watchmen who would guard the gates of the town, and when needed, raise a hue and cry. This early police force had an additional special function. They were to assure that vagabonds, vagrants, and other people acting to upset the peace did not find shelter. Sometimes, they would even be watchdogs at farms, wake up the workers in the morning, make sure they worked efficiently, and time the lunch breaks. The constable became a "neutral" agent between disputants. The Statute of Winchester provided for fines for those who refused to serve as parish constables.

In the Peace Act of 1361, the English crown recognized the relationship between a constable and a justice of the peace. A layer of justices was established, over the parish constables, who were officials not of the parish but of the county or municipal government, and it was this system that endured until organization of the professional police in 1829.

Between the fifteenth and eighteenth centuries, the office of the peace and the position of the parish constable underwent a devastating transformation. Political conflicts and corruption that led to politically convenient appointments finally resulted in complete deterioration of this function. Most commonly, the person whose turn it was to serve would hire a stand-in. Typically, the men hired had no qualifications other than a willingness to work cheaply. Frequently, they were physically or mentally challenged; often, the village idiot served as parish constable (Klockars, 1985).

Preventive Policing

With the growth of cities, the more sophisticated criminality, and more police corruption, law enforcement had become highly ineffective in the eighteenth century. Patrick Colquhoun (pronounced "Cohoon") understood the need for change, and in 1792, under the police act of that year, he was appointed a justice of the peace in London. This branch of the government had the goal of reforming the disreputable and corrupt magistracy of the metropolis. Colquhoun brought both compassion and a deep sense of social responsibility to this position, and he had many successes, such as enforcing a law against embezzlement and capturing counterfeit money.

Patrick Colquhoun understood that a police force had to be preventive in nature, and wrote two important books on the topic. In *A Treatise on the Police of the Metropolis*, he emphasized that the policing in England should be considered a new science and he described police as "all those regulations in a country which apply to the comfort, convenience and safety of the inhabitants" (Colquhoun, 1969 [1800]). Colquhoun wanted to improve the system and believed that this could be done by coordination and direction at the national level. This would come from a board consisting of commissioners who each would have his own magistrates in different parts of the country (England). The magistrates would then have constables under them who would supervise the petty constables.

Colquhoun had high hopes, and his second book, *A Treatise on the Commerce and Police of the River Thames* (1969 [1806]), is a documented story of how the first preventive police force was instituted in England. At the end of the eighteenth century the water around London was the busiest in the world, and it was also a place of much crime. Criminality had become a part of daily life, and Colquhoun had a plan to minimize the crime.

In 1798, a marine police force was established, consisting of four departments: judicial department, preventive department, department for employing labor to discharge ships, and prosecution department. Colquhoun became the superintending magistrate. In addition, there were a resident magistrate, law clerks, chief constables, petty constables, ship constables, and some other labor. Their success was spectacular, and most crime vanished. Statistics from this period show that theft was reduced by 95 percent in eight months. At the end of his book, Colquhoun describes the power of the police as follows: "General Prevention—mild in its operations, effective in its results; having justice and humanity for its basis, and the general security of the State and Individuals for its ultimate object" (Colquhoun, 1969 [1800]). The government took over full responsibility for the marine police in 1800 and extended the operations of the police force while keeping Colquhoun's principles intact.

ORIGINS OF MODERN POLICING: FRANCE

The French police system that is in use today has served as a model for several countries in Europe and around the world, in particular those countries in which France had a hand in the government. The police model goes back to the

Napoleonic era. The emperor was the essential factor in enabling reconstruction of the law enforcement agency of the Ancient Regime. Reconstruction occurred in the late eighteenth century after the revolution. By the turn of the century Napoleon had established a structured police force. The main features of this new agency were control from the center by the minister and his prefects, a dual structure of military and civil elements, a distinct police force for the capital city, substantial reserves for emergency use, and a high degree of surveillance of the population (Stead, 1977).

In the first three and a half years of existence, the Napoleonic police had nine different ministers of police. However, in 1799 the most powerful minister was appointed, Joseph Fouche, who would occupy the post for the next 16 years. Together, Napoleon and Fouche built the most effective police force in the world at that time, and they made Paris safer for its citizens than were London, Rome, or New York. The basic structure of this system remains in the police force of France today (Stead, 1957).

Prior to the revolution, during the Ancient Regime and until the end of the Middle Ages, the king had a special army that had an internal police element. Its basic function was protection of the king's personnel in the rural areas and along the main roads. In the middle to late seventeenth century, the police became more important and duties increased. For example, the king had a new guard force created for him, a police force was instituted for the purpose of fighting violent outbursts, and police officials of magisterial status were introduced. In the early eighteenth century the *marechaussée*, a royal army that functioned as a bodyguard for the king, was added to the police force (Stead, 1983).

After the revolution in 1789, the police force was associated primarily with bad memories, and the changes instituted by Napoleon and Fouche were necessary. Overall, the police of France have existed as a professional body longer than has any other police agency in Europe. Nevertheless, they resemble the U.S. and British police in being under the rule of law and ultimately answerability to law and to a democratically elected legislature.

LAW ENFORCEMENT IN THE NINETEENTH CENTURY

Throughout the nineteenth century, London and New York were growing into major cities. London doubled its population between 1750 and 1820, and New York quintupled its population between 1790 and 1830. As a result, many new immigrants were forced to live in slums and criminality increased drastically.

Ironically, a moral reformation took place in both England and the United States in the midst of the growing criminality. The harsh and physical penalties were replaced by imprisonment and an effort to change the methods of punishment. The moral was that an offender should have the chance to change his lifestyle. In addition, social and legal reforms that occurred around this time resulted in more reports of complaints and crimes (Miller, 1977).

Public opposition to the establishment of a police force in England stemmed from the perception that long-cherished liberties would be swept away, a perception based partially on the history of public police on the continent, more specifically, the French police. The public police in France was associated with a reign of terror and oppression. However, Sir Robert Peel skillfully managed to convince the British Parliament about his proposal to establish a new municipal police force in 1829. Historians view Peel as a master of facts and arguments, who promised the Parliament (and ultimately the public) slow, experimental reforms but did not reveal in his speech the drastic changes he really contemplated. He "sold" his police as mainly a preventive force; however, soon after its establishment it was used for other purposes as well, during riots and other public dissents, playing precisely to the fears that opposed its creation.

Several riots threatened the ruling minority in the early nineteenth century, but Peel's Metropolitan Police Force quieted most of the wealthy upper-class outcry. The story was different for the middle and working classes. Secret meetings were held and men were trained to use violence to force compliance with their demands. The tension increased and in 1848 a riot broke out in which the working class was eventually defeated. Tensions remained high for the next 20 years, but a resolution was reached in 1867, when workers finally received the right to vote. During the various crises in London, the police force always had the difficult task of defending a social order that was maintained by a ruling minority, and they were therefore seen as supporters of only the wealthy class up until 1867.

LAW ENFORCEMENT IN COLONIAL AMERICA

American policing can be traced back to the arrival of the earliest colonists. They carried with them to the New World ancient Saxon roots and seventeenth-century English police traditions. At first, colonists used a mutual protection system in which they looked out for each other. However, soon they began to use town constables, sheriffs, and night watchmen. In the beginning, the majority of settlers went to either the northeast or the south. In the north, more towns and cities developed, and as a result, a patrolling force of watchmen became the system of choice for new settlements. In the flat and fertile south, the more agrarian settlers brought with them law enforcement styles that had grown up in the countryside of England. This included the county sheriff, the earliest of which may have appeared around 1634 in Virginia. Eventually, however, the U.S. sheriff became the most powerful and significant law enforcement figure in the southern United States. Western areas were settled later than northeastern and southern areas, but when they were, the sheriff was adopted as the primary police agent. Among early sheriffs were Charles Lynch (known for his method of executing criminals, "lynching"), "Wild Bill" Hickock, Wyatt Earp, and Pat Garrett.

Industrialization in the north resulted in the growth of bigger towns and cities. As the sheriff system did not appear to be a sufficient law enforcement method for these areas, the night watchmen system was brought in from England. The primary task was to prevent crime, but in addition, members were charged with lighting street lamps and keeping an eye open for fires. They would sometimes chase runaway animals or assist in family disputes (Moore and Kelling, 1985).

The principal law enforcement strategy practiced by early watchmen was to raise a hue and cry if they saw a crime being committed. To serve as a watchman was a civic responsibility. Boston organized the first watchmen system in 1631. It was organized within a military structure with one commanding officer and six watchmen. The system was later replaced by a volunteer citizen patrol in which people who did not volunteer were fined a stiff penalty (Lane, 1967).

New York instituted a night watch patrol in 1658 (New York was at that time named New Amsterdam). This force became known as the *rattle force*, because the eight-man brigade carried wooden rattles to announce their presence. The idea was to scare a lawbreaker before a confrontation could occur, and in addition, the wooden rattles allowed members to communicate with each other. This civic duty was a paid responsibility imposed in turn on every citizen. People who did not do their duty were fined.

In 1700, Philadelphia introduced its own night watch system. In 1712, Boston created the first paid full-time night watch force. The men received approximately 50 cents a night (Bobb and Schultz, 1972). Cincinnati and New Orleans followed in 1803 and 1804, respectively.

ORIGINS OF MODERN POLICING: THE UNITED STATES

Colonial America was characterized by mob activity, and at the middle of the nineteenth century, there was no national detective agency in the United States, and only Texas had a statewide police force. The modern police department came late to the United States. It seemed a contrast to the American ideals that a uniformed corps, controlled by politicians, should patrol the streets. The old system was based on citizen involvement. There would be one or a couple of people who would be in charge, but citizens would be brought in when help was needed. This changed with the constable-nightwatch system (Fosdick, 1969), which lasted for the first half-century of America's existence. Constables, elected from the male inhabitants of a city, served as retrievers of stolen goods and preventers of crime and fire. The constable-nightwatch system focused mostly on late evening crime, and the system was finally abandoned in the mid-1840s. U.S. cities were increasing rapidly at this time, and daytime police were needed. As a result, a police department was established that focused on preventive policing during both day and night.

A municipal police force developed differently in New York from London. New York was not the nation's center at the turn of the nineteenth century. The Civil War was still going on and most political issues were discussed in Washington, DC. However, the enormous increase in immigrants, and all the problems that came with them, made a police force a necessity in New York. In addition, several riots occurred in 1834. These included the Stonecutter's Riot, Five Points Riot, O'Connell Guard Riots, and Chatham Street Riots (Costello, 1972 [1885]). As a result, media and citizens started campaigning for a police system along lines similar to those of the Metropolitan Police Force. This finally occurred in 1845, and the task for the New York City police of the nineteenth century came to uphold the political order of a representative government.

However, this permitted mistreatment of certain immigrant populations in the city; the Irish population, especially, became a target. In addition, the police could not keep up with the criminality, and as a result, private policing agencies began to flourish in the mid-nineteenth century.

New York became the first city in the United States to introduce a modern police department. Cincinnati and New Orleans followed the New York standard in 1852. Boston and Philadelphia came after, in 1854, Chicago followed the year after, and Baltimore in 1857 (Ketcham, 1967). However, there were still several problems that the new police had to overcome. The economic depression in the mid-1850s resulted in increased crime. There were also public suspicion and distrust of the police, a residue of feeling from the past. In addition, early police departments also lacked uniforms. Because the newly developed police department lacked these necessities, private police organizations were established in some cities.

In Chicago, the infamous Allan Pinkerton started the Pinkerton Protective Patrol (Morn, 1977). This police force had the same functions as those of the regular police department. Pinkerton expanded his offices after the Civil War and opened up in New York City (1865) and Philadelphia (1866). However, his officers focused on detection work, not patrolling and prevention. The agency was successful, and by 1895 nine offices were established in various U.S. cities. By this time the public police had developed new detective techniques, and in 1893 the National Association of Chiefs of Police had been established. As a result, by the turn of the century the public police had taken over most detecting and protecting work.

Southern Slave Patrols

The slave patrol forces, which were established in the southern states in the early eighteenth century, can be characterized as a direct forerunner to modern policing. Their main function was patrolling and policing, in contrast to the existing constable, watchman, and sheriff, who had as additional duties acting as fire watchers and/or to collect taxes. The slave patrol can be seen as a continuous development of *informal policing* (Lundman, 1980), characterized by community members sharing responsibility for maintaining order. This patrol force kept some of the original networks, but it also included modern police offices and procedures. As a result, the slave patrol can be referred to as *transitional policing* (Lundman, 1980).

There are three main reasons why slaves constituted a threat to the southern states. The greatest problem was slaves running away from their owners. Criminal acts such as theft, robbery, crop destruction, and arson constituted another major problem. Finally, conspiracies and revolts were a third form of threat.

Originally, any person could apprehend, chase, send home, kill, or destroy slaves who acted out or committed criminal acts. Later, with societal changes, these informal means became inadequate and slave patrols werre made a part of the colonial militias. These militias had multiple tasks, and it was not until the early eighteenth century that the slave patrols became a separate entity within the militias.

The colony of Carolina was the first to recognize a slave patrol force in 1704. The white population was afraid of uproars because of the high number of slaves, and therefore a force was created whose main task was to control the slaves. However, this unit only lasted a short time, and not until 1734 was a regular slave patrol established. Other states, including Tennessee, Louisiana, Arkansas, Georgia, Mississippi, and Missouri, followed, and by the middle of the nineteenth century most of the southern states had slave patrols.

The number of men on patrol was limited. Usually, between 5 and 15 men would make weekly rounds. Any white person between 16 and 60 years of age could be called in for duty. Each district had three patrol commissioners, who were responsible for the patrol lists. In addition, there was one captain for each district who was responsible for the men. The pay varied from salaries and rewards to exemption from other duties.

The slave patrol had a number of limitations. The elite members of the various districts often avoided duty by paying a fine or finding a substitute. Inappropriate behavior among the patrol force was often reported. Sometimes the patrollers would drink too much liquor before or during duty. Other times, the punishment of the slaves was too harsh. In addition, training was infrequent and often very poor. The slave patrols disappeared with the freeing of the slaves (Reichel, 1988).

Policing in the United States in the Twentieth Century

In general terms, policing in the twentieth century is divided into three eras or periods: political, professional, and community oriented. A wealth of information exists about these; however, the main purpose here is just to sketch the critical developments that underline the theme of who the police served during these eras rather than give a detailed description of all the developments during these periods.

Walker (1999) argues that policing in the United States in the nineteenth century was characterized by an epidemic of corruption. The police took payoffs for not enforcing laws on drinking, gambling, and prostitution. Officers often had to pay bribes for promotion. Corruption was one of the main functions of local government, and the police were only one problem. Political reformers made police corruption a major issue during the nineteenth century; however, in general the success was only limited.

From about the middle of the eighteenth century to the 1920s, local policing was dominated by politics; this era was named *the political model* of policing—oriented to special interests. Politics influenced every aspect of law enforcement during this period, from employment, to promotion, to appointment of the police commissioner or chief of police, to some police arrest practices and services—all were determined by political considerations. Police jobs became an important part of the political patronage system that developed in the cities. The police were particularly useful during elections because they maintained order at polling booths and were able to determine who voted and who did not (Roberg et al., 2000). The amount of law enforcement received, if any, was dependent on one's political connections (Walker, 1977).

The second period, the *professional era* of policing, began in the early years of the twentieth century but was formally defined as the period between 1920 and the late 1970s (the exact definition varies with different approaches and is sometimes ended at the 1960s). The popular term "get politics out of the police and get the police out of politics" serves as a basis for the call to hire professional administrators and so limit political influence. This period is associated with the names Richard Sylvester, the founder of the International Association of Chiefs of Police; August Vollmer, police chief of Berkley, California, founder of the American Society of Criminology and avid advocate of higher education and extensive training for police officers; and Orlando W. Wilson, author of one of the most influential texts in police science, *Police Administration*.

Although the reformers were all powerful and highly influential men, their success in getting the politics out of the police and the police out of politics was limited. The downside to the professionalization of the police force was its removal from beat functions and close interaction with the public and seclusion and isolation inside a police car. The political turbulence that dominated the United States in the 1960s once again positioned the police on the side of the politicians. War protests, the civil rights movement, street crime, union actions, and unrest on college campuses created problems for city police forces (Hunter et al., 2000). The need to represent the "oppressive" government, coupled with aggressive techniques of crowd control and the discovery of rampant corruption in the New York City Police Department, once again highlighted the role that police play in the society—and this role was not primarily to serve and protect the public.

SUMMARY

The antecedents of the third and final era of policing, the community policing era, can be traced to the new wave of reforms, led by New York City Police Commissioner Patrick Murphy during the 1960s and 1970s. Service and protection of the public based on crime prevention and close cooperation with the public became the main objective of the new philosophy. A number of attempts were made to restore and enhance police–community relations, through experimentation with the way that police services were delivered. Concepts such as team policing, Integrated Criminal Apprehension programs, neighborhood foot patrols, Problem-Oriented Policing, Community-Oriented Policing, and COPPS (Community-oriented policing and problem-solving) became central to the success of the new approach. A number of highly publicized research studies were undertaken, all in an attempt to find the right formula to enhance the delivery of services. For the first time in the history of policing, a *theoretical* change in orientation took place—the police's function was to serve, protect, understand, and cooperate with the community. The word *theoretical* is emphasized because changes in police training are minimal if not ignored completely. Despite denial on the part of politicians and some high-profile police chiefs, the resources dedicated to police training remain inadequate and poorly utilized. The change in orientation is pronounced but the implementation is minimal—politics are still deeply engrained in the twenty-first-century policing, and the police are still highly involved in politics. The proof? Let us take a closer look at how and by whom police officers are trained today versus what is recommended, needed, and desired. Hopefully, by the end of this book the reader will realize how dangerously close we still are to the qualifications of the medieval village idiot.

References

Berg, B. L. (1992). *Law Enforcement: An Introduction to Police in Society*. Needham Heights, MA: Allyn & Bacon.

Bittner, E. (1970). *The Function of the Police in Modern Society*. Washington, DC: U.S. Government Printing Office.

Bopp, W. J., and Schultz, D. O. (1972). *A Short History of American Law Enforcement*. Springfield, IL: Charles C. Thomas.

Colquhoun, P. (1969 [1800]). *A Treatise on the Police of the Metropolis*, 7th ed. Montclair, NJ: Patterson Smith.

Colquhoun, P. (1969 [1806]). *A Treatise on the Commerce and Police of the River Thames*. Montclair, NJ: Patterson Smith.

Costello, A. E. (1972 [1885]). *Our Police Protectors: A History of New York Police*. Montclair, NJ: Patterson Smith.

Davies, R. W. (1977). Augustus Caesar: a police system in the ancient world. In P. J. Stead (ed.), *Pioneers in Policing*. Montclair, NJ: Patterson Smith.

Echols, E. (1957). The Roman city police: origin and development. *Classical Journal, 53*, 377–385.

Echols, E. (1961). The provincial urban cohorts. *Classical Journal, 57*, 25–28.

Engels, F. (1942). *The Origin of the Family, Private Property and the State*. New York: International Publishers.

Fosdick, R. B. (1969 [1920]). *American Police Systems*. Montclair, NJ: Patterson Smith.

Hunter, R. D., Mayhall, P. D., and Barker, T. (2000). *Police–Community Relations and the Administration of Justice*. Upper Saddle River, NJ: Prentice Hall.

Ketcham, G. A. (1967). Municipal police reform: a comparative study of law enforcement in Cincinnati, Chicago, New Orleans, New York and St. Louis, 1844–1877. Ph.D. dissertation. University of Missouri.

Klockars, C. (1985). *The Idea of Police*. Thousand Oaks, CA: Sage Publications.

Lane, R. (1967). *Policing the City: Boston, 1822–1885*. Cambridge, MA: Harvard University Press.

Lundman, R. J. (1980). *Police and Policing: An Introduction*. New York: Holt, Rinehart and Winston.

Mair, L. (1964). *Primitive Government*. Baltimore: Penguin Books.

Miller, W. R. (1977). *Cops and Bobbies*. Chicago: University of Chicago Press.

Moore, M. H., and Kelling, G. L. (1985). To serve and protect: learning from police history. In A. S. Blumberg, and E. Niederhoffer (eds.), *The Ambivalent Force*. New York: Holt, Rinehart and Winston.

Morn, F. T. (1977). *Allan Pinkerton: Private Police Influence on Police Development*. Montclair, NJ: Patterson Smith.

Reichel, P. L. (1988). Southern slave patrols as a transitional police type. *American Journal of Police, 7*, 51–77.

Reith, C. (1975). *The Blind Eye of History: A Study of the Origins of the Present Police Era*. Montclair, NJ: Patterson Smith.

Reynolds, P. K. (1926). *The Vigiles of Imperial Rome*. London.

Roberg, R., J. Crank, and Kuykendall, J. (2000). *Police and Society*, 2nd ed. Los Angeles, CA: Roxbury Publishing Company.

Sacks, K. (1979). *Sisters and Wives: The Past and Future of Sexual Equality*. Westport, CT: Greenwood Press.

Schmalleger, F. (1993). *Criminal Justice Today*, 2nd ed. Upper Saddle River, NJ: Regents/Prentice Hall.

Stead, P. J. (1957). *The Police of Paris*. Rochester, Kent, England: Stables Printers.

Stead, P. J. (1977). Joseph Fouche: the Napoleonic model of police. In P. J. Stead (ed.), *Pioneers in Policing*. Montclair, NJ: Patterson Smith.

Stead, P. J. (1983). *The Police of France*. New York: Macmillan.

Walker, S. (1977). *A Critical History of Police Reform*. Lexington, MA: Lexington Books.

Walker, S. (1999). *The Police in America: An Introduction*, 3rd ed. New York: McGraw-Hill.

TRAINING VERSUS EDUCATION
Conceptual Framework

INTRODUCTION

This chapter is based on materials gathered from many police departments around the United States and other countries (especially the British Police probationer training program and the Canadian Police training models). Experiences of police departments varying in size, resources, politics, and geographical locations contributed to eclectic and diversified views on the most appropriate methods of preparing police officers for their jobs. None of the methods or techniques here is the best or most appropriate one, but every law enforcement organization, regardless of its size and resources, will be able to find helpful insights, tips, and approaches by identifying the most appropriate techniques and methods or by using a piecemeal approach.

EDUCATION VERSUS TRAINING

Education involves the learning of general concepts, terms, policies, practices, and theories. The subject matter taught is often broad in scope. Not only are typical relationships and practices within a given field discussed, but so are hypotheses as to why these particular relationships and practices exist. Among the types of skills stressed in education are analyzing different situations correctly; both communicating information and defending one's opinion effectively orally and in writing; drawing insights from related situations in different settings; gathering information using various methods of research; creating alternative approaches and solutions to diverse problems; and learning new facts and ideas from others though various media (e.g., lectures, books, articles, conversations, video presentations). The goals of education include teaching people to recognize, categorize, evaluate, and understand different types of phenomena; to interact and communicate effectively with others; to think for themselves; and to predict the probable outcomes of competing solutions.

The goal of training is to teach a specific method of performing a task or responding to a given situation. The subject matter taught is usually narrow in scope. Training usually involves two stages:

1. Prescribed procedures are presented and explained.
2. The procedures are practiced until they become second nature or reflexive.

Training is focused on how most effectively to accomplish a task whenever a particular situation arises. Training is experiential and goal oriented.

Among the skills associated with most training programs are the ability to determine whether or not the circumstances warrant following a prescribed course of action, the physical and verbal skills associated with those actions, and the cognitive abilities needed to recall what steps should be followed and in what order for each of the situations covered in the training program (Timm and Christian, 1991). Timm and Christian (1991) identified the advantages and disadvantages of training versus education in terms of the way they apply to the environment of private security. For the purposes of this book, their concepts were converted into police training environments. The advantage of training and education are as follows:

Training	Education
Training prepares a person with a ready response in case of emergency.	Skills can be applied to various situations.
"Programmed" responses can be attained through intensive training.	Education results in a wider range of knowledge, and more intelligent communication skills.
Research is used to determine the best responses.	Education provides knowledge of how to create good training programs.
Training makes people feel more confident.	Education may result in more worldly knowledge and thereby more tolerance of differences.
Training leads to quicker and more efficient responses.	Education takes the student through an extensive program that prepares him for a wide range of occupations.
Training leads to more consistent responses that are in accordance with the authority.	Education provides greater awareness of contemporary and historical events.
The training process is concentrated and inexpensive.	Education provides people with better logical solutions.

Training	**Education**
Skills that require hands-on training are acquired efficiently.	Education provides problem-solving skills, critical thinking, and communication skills.
Training provides an alternative solution to people who do not have the interest or ability to find their own solution.	
Training decreases the likelihood of being sued because of the appropriate training to specific situations.	

Disadvantages are:

Training	**Education**
Training is situation-specific, and no two situations are the same.	Education is often expensive and has a diffuse focus.
It can be difficult to improvise a solution if the problem differs from training.	Education does not provide specific technical training.
Correct responses tend to change more often than appropriate training.	Programs are long and people may not have the interest to complete them.
Training eliminates creativity in responses.	Programs offer no "pat" answers, which can be frustrating.
Training may result in people who are unhappy with the responses the training provides.	

THE ROLE OF TRAINING AND EDUCATION IN LAW ENFORCEMENT

Both training and education play important roles in the field of law enforcement. Training provides officers with unambiguous instructions on how to perform many of the tasks that they are expected to complete. As an outcome, trained officers often respond both more consistently, using proven techniques, and more automatically, even under emergency conditions. Education, in contrast, helps prepare officers to solve problems independently as well as to communicate and interact effectively with others.

Different law enforcement positions may require different levels of education and training; however, a combination of both is needed in every position. Law enforcement officers, for example, often interact with people from a wide range of backgrounds and exercise considerable discretion in many critical situations (e.g., deciding whether or not to arrest someone, whether or not to shoot, whether or not to evacuate an area in an emergency situation) and have to prepare written incident reports. These tasks can be performed effectively only if through acquisition of some general skills traditionally offered through various educational programs. The officers also need training in a wide range of specific tasks directly related to their positions (e.g., arresting people, shooting firearms, operating equipment, and handling emergency situations).

Middle- and top-level administrators also need training in certain areas, even though in performing most of their tasks they rely more heavily on knowledge and skills generally acquired through formal education. For example, middle-level and senior police executives often need training in how to operate computers, use new software (e.g., crime mapping software) and other technologies, decide what reporting procedures to follow, and handle a number of other essentials that will enable them to complete the tasks for which they are responsible. Law enforcement executives may also participate in training programs to familiarize themselves with new evaluation tools and research findings. Both training and education appear to be essential regardless of the law enforcement position one holds within an organization (Timm and Christian, 1991).

MERGING TRAINING AND EDUCATION

Canadian Police Education and Training Models

The Canadian police currently have four models of basic training for police recruits.

Model 1: Separation of Police Education and Training from Mainstream Adult Education.   Recruits and police officers who have at least a grade 12 level of education are trained in institutions that are separate and independent. One such recruit training program, for the Ontario Police College, includes six levels of training, starting with level I, composed of field training by the individual police agency; followed by level II, with classroom training; level III, field training; level IV, classroom training; level V, general duties; and finally, level VI, optional specialized training at the college.

Model 2: Model 1 Delivered on a University Campus.   In model 2, police staff teach classes on police administration and procedures; academic staff teach classes on criminal justice and social services; and lawyers and judges teach classes on criminal law, evidence, and procedures.

Model 3: Holistic Approach to Recruit Training.   The holistic approach is based on the assumption that police recruits should be exposed to the entire criminal justice system rather than just to the field of policing. The basis of this approach is a block training program, which alternates classroom learning with field experience. This five-block three-year training program consists of block I, basic police training at the police academy; block II, field training; and block III, further work at the academy. Upon successful completion, recruits are able to work alone when they return to their departments; however, they return to the academy for blocks IV and V.

Model 4: The Quebec Model. The Quebec model attempts to integrate police education with adult education. New recruits are required to complete a three-year college program to obtain a Diploma of Collegial Studies, which includes general academic courses and instruction in criminology, policing, and law. Following this, candidates are sent to the Quebec Police Institute in Nicolet. The training center places a strong emphasis on "hard" police skills, including driver training, firearms, and arrest and control techniques, as well as "soft" skills, such as physical fitness and community relations. The intent of the training center is to break knowledge down into a variety of disciplines than can be enhanced when the recruit begins his or her field training on the street (Griffits et al., 1999).

British Police Education and Training Model

While revising their basic police training in 1999, the British Police Training Center identified a number of theories and models related to learning concepts. These theories are summarized in Tables 3–1 to 3–4 and their applicability to various stages of police training can be identified by individual trainers in their respective training environments.

Training Concepts

In its formal sense, training is a consciously selected means to a particular end. More and Wegener (1996) identified a number of training techniques that managers use to achieve specific objectives. The techniques can be divided into five categories (pp. 423–424):

1. *Orientation.* The academy's basic training curriculum is designed to give rookies an orientation to real police work as opposed to the "Dirty Harry" stereotype.
2. *Indoctrination.* A deliberate attempt is made to internalize acceptable perspectives, attitudes, norms, and values in police trainees, on the assumption that they will become internal control mechanisms, capable of regulating a trainee's personal conduct and job-related behavior.

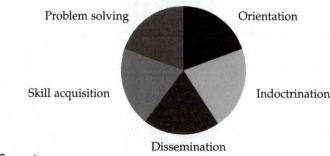

Training Concepts

Table 3–1   Learning Theories

Theory	Developed by:	Main Concept
Classical conditioning	Pavlov (1927), Watson (1930)	Existing responses come under control of a new stimulus. There is an association between stimulus and response.
Operant conditioning	Skinner (1965)	Existing behavior is modified through positive reinforcement.
Negative reinforcement and punishment	Heckel et al. (1962)	Inappropriate behavior is punished to prevent the behavior from reoccurring.
Social learning theory	Bandura (1977)	Behavior is learned by imitating and observing other people.
Locus of control theory	Rotter (1966)	This is a social learning theory in which it is believed that people can influence their actions and the outcome (internal locus of control) directly or the actions control the people and the outcome (external locus of control).
Field cognition theory	Tolman (1951)	Cognitive maps are created in the mind as a result of experience, and this map increases when new experiences are learned.
Self-directed learning	Taylor (1979)	Learning is a circular process in which four phases and four transitions have to be fulfilled:

1. *Disconformation (transition):* a major discrepancy between expectation and experience.
2. *Disorientation:* a period of disorientation and confusion.
3. *Naming the problem (transition):* naming the problem without blaming self or others.
4. *Exploration:* gathering insights, confidence, and satisfaction.
5. *Reflection (transition):* a private reflective review.
6. *Reorientation:* a synthesis experience that results in a new approach to the learning task.
7. *Sharing the discovery (transition):* testing the new understanding.
8. *Equilibrium:* the new approach is elaborated, refined, and applied.

Table 3–2   Learner Groupings

Grouping	Developed by:	Main Concept
Individual learner	Jacobs and Fuhrman (1984)	Learning can take place individually. The person is comfortable working alone and is determined to learn. Time is essential because this form of learning may be lengthy.
Groups of learners	Zajonc (1965)	If individual learning has taken place prior to group learning, the group learning will be more effective.
	Tuckman and Jensen (1977)	Effective group learning occurs after the group has gone through five stages together: forming, storming, norming, performing, and adjourning.
Learner proximity	Sommer (1967)	Interaction between group and learner, such as class discussions, and a horse-shoe classroom setting that provides better learning.

Table 3–3   Learner Motivation

Motivation	Developed by:	Main Concept
Expectancy-valance theory	Tolman (1932)	Motivation is based on four variables: *expectancy* refers to the relationship between effort and outcomes; *outcomes* are the consequences of behavior; *instrumentality* relates outcomes to rewards; and *valance* is the outcome. In essence the theory postulates that learning is easier when the individual believes that he or she will succeed.
Hierarchy of human needs	Maslow (1943)	There is a hierarchy of needs, ranging from primitive to mature. A person has to fulfill one level before he or she can move up to the next one.
Motivators and hygiene factors (two-factor theory)	Herzberg (1959)	Two factors are involved in learning: *Motivation* refers to the elements of what the person is doing, such as interests, challenge, and responsibility; *hygiene* reflects such needs as the quality of the physical environment and safety.
Cognitive dissonance	Festinger (1977)	A conflict between two ideas results in elimination of the idea that does not result in consistency, ensuring that the idea most beneficial to the learner is used.

Table 3–4 Learner's Neurolinguistic Programming

Neurolinguistic Programming	Developed by:	Main Concept
Learner's perceptions	Dilts and Mayer-Anderson (1980)	Learning is achieved through primary factors (rapport and communication) and main factors (visual, auditory, and kinesthetic).
Social styles	Merrill and Reid (1981)	There are four basic states related to how to interact with others: driver, expressive, analytical and amiable. All are observable in childhood and therefore can be manipulated early.
Learner's styles	Kolb (1976)	There are four categories of learning style: behavioral, structural, functional, and humanistic.
Humanistic theory	Rogers (1951)	Learning is best when it involves the person as a whole, on both emotional and intellectual levels.
Double-focus theory	Cohen (1972)	Learning achievement is related to how well a teacher manages to balance theme and process in the topic that is being taught.
Gestalt theory	Perls (1969)	This theory is based on a humanistic approach in which a person has to be aware of and accept his or her feelings in order to function as a whole, after which effective learning can take place.
Learning from simple to complex	Skemp (1962)	Schemata are developed in forms of experiences that help form and give meaning to future learning.

- Police instructors teach the "right way" to do things.
- Conformity is rewarded.
- The negative consequences of deviant behavior are emphasized.

3. *Dissemination.* This is the intentional and structural communication of specialized information needed by a trainee to function as an efficient, effective, and productive member of the department or specialized unit: for example, violent death investigation seminar for patrol officers designed to provide trainees with information relevant to their critical role in the preliminary investigation.

4. *Skill acquisition.* This is intended to give trainees the opportunity to master a special skill through simulation or repetition: for example, combat shooting on the fire range.

5. *Problem solving.* A technique is that that uses comparative analysis in an attempt to make decisions that fall within the range of solutions acceptable to management.

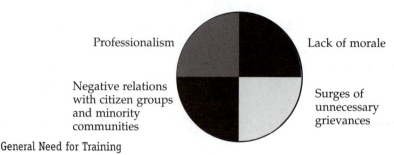

Professionalism

Lack of morale

Negative relations
with citizen groups
and minority
communities

Surges of
unnecessary
grievances

General Need for Training

Need for Training

The general need for training in law enforcement environments can be represented in terms of four major problems:

1. Lack of morale
2. Surges of unnecessary grievances
3. Negative relations with citizen groups and minority communities
4. Lack of professionalism

Solutions to these four concerns are an integral part of high-quality policing. An officer who lacks morale due to inadequate knowledge of the basic tools required to perform effectively will eventually perform in a way that is less than professional. As a result, surges of unnecessary grievances will flood the organization that he or she serves, followed by negative relations with citizen groups and minority communities. This, in turn, will affect the professional image of policing, which in a vicious circle will again lead to a decrease in morale.

Whisenand and Rush (1993) have identified six philosophical planks that provide a foundation for modern police training. While giving scrupulous attention to a particular training program, it is also essential to consider these philosophical approaches prior to a final commitment to a given curriculum. Training content can be relevant and reflect the political, social, and professional stance of a given agency, but inattention to more complex considerations may lead to a lack of effectiveness in its implementation.

MODERN POLICE TRAINING

Philosophical Foundation

1. Motivation plus acquired skills lead to positive action.
2. Learning is a complex phenomenon that depends on:
   - The motivation and capacity of each person
   - The norms of the training group

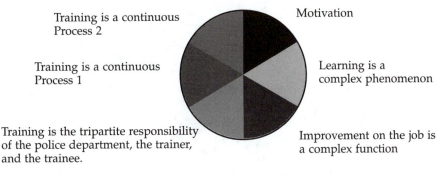

Training is a continuous Process 2

Motivation

Training is a continuous Process 1

Learning is a complex phenomenon

Training is the tripartite responsibility of the police department, the trainer, and the trainee.

Improvement on the job is a complex function

Six Philosophical Planks that Provide a Foundation for Modern Police Training

- The institutional methods and behavior of the trainers
- The climate of the police department
3. Improvement on the job is a complex function involving such factors as:
    - Individual learning
    - The shared expectations of the workforce
    - The general climate of the department
4. Training is the tripartite responsibility of:
    - The police department
    - The trainer
    - The trainee
5. Training is a continuous process that serves as a vehicle for constant updating of:
    - Knowledge
    - Attitudes
    - Skills of the department's human resources
6. Training is a continuous process and method for consistent improvement in the capacity of individual officers to act as a team.
    - Training is never fully accomplished.
    - Training is always in process.
    - Training is designed to focus knowledge so that it can be applied in specific situations. Department-sponsored police training has two very basic goals:
        (1) To improve an officer's on-the-job performance.
        (2) To develop an officer's capacity to handle even higher levels of responsibility.
    - Training should help the individual to do a better job, while preparing the person for even more challenging duties.
    - Training is unique and promotes both challenge and unity of purpose within an organization.

Conducting and evaluating
the training

Identifying training needs

Preparing the
training program

Preparing training objectives

The Training Cycle

Training Cycle

The following training cycle is proposed by Plunkett (1992):

1. Identifying training needs
   - Scan the environment.
   - Identify when performance is not meeting expectations or standards.
   - Compare performance to standards on a regular basis.
   - Recognize that training becomes necessary when and where significant differences exist between performance and standards.
   - Translate needs into a list of tasks to be taught.
2. Preparing training objectives
   - Determine the training criteria.
   - Trainer and trainee must share common objectives.
   - State in writing the expectations, conditions, and tasks to be performed.
   - Determine the standards of performance.
3. Preparing the training program
   - Who is to do the training?
   - Who is to be trained?
   - When will the training take place, and how much time will be set aside?
   - Where will the training take place?
   - What equipment is needed?
   - How and in what chronological order will tasks be taught?
   - What specific methods of instruction will be used?
   - How much money and/or what other resources will be required to ensure a successful training effort?
4. Conducting and evaluating training
   - Provide immediate feedback once execution begins.
   - Assess the effectiveness frequently.
   - Re-teach if necessary.

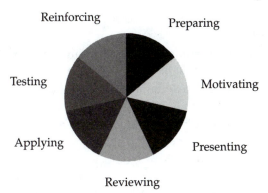

The Instructional Protocol: The Heart of the Teaching Process

Instructional Protocol: The Heart of the Teaching Process

Regardless of the method selected, good training usually adopts an instructional protocol that includes the following actions (More and Wegener, 1996):

1. *Preparing.* Select the right topics for the trainees and their needs.
2. *Motivating.* Get attention, promote interest, create desire ("you tell me how").
3. *Presenting.* Use interactive techniques and multimedia.
4. *Reviewing.* Summarize, review, and conclude.
5. *Applying.* Blend theory with practice.
6. *Testing.* Evaluate knowledge through testing.
7. *Reinforcing.* Provide positive support for proper performance.

Common Deficiencies in Current Training Programs

Thibault et al. (1998) identified a number of common deficiencies in training programs offered to police officers around the country (pp. 315–316).

1. *Program content.* This should include human relations, communications, adult versus juvenile behavior, and other topics.
2. *Quality control of instructors.* Training personnel pulled from the field often do not know how to teach (they tell "war stories" rather than acting as educators). Can such instructors communicate effectively and serve as role models for recruits?
3. *Training facilities.* Are facilities state of the art or simply former military barracks, jails, and the like? What is the impact of the training environment on successful retention of material?
4. *Training equipment.* Teaching should not only be verbal: 7 hours of lectures 5 days a week = 35 hours of boredom!

5. *Part-time personnel.* Instructors and facility directors often perform training duties only on a part-time basis.

6. *Full-time attendance.* Part-time students and part time workers make for tired officers and half-trained students who fall asleep during instruction.

7. *Training before power.* Training is mandatory before granting power to new police officers.

8. *Follow-up evaluation.* Follow up at planned intervals for at least for two years after graduation.

9. *Field training officers and programs.* Where should we start?

More and Wegener (1996) add additional dimensions to the current training deficiencies by identifying 10 cardinal errors to be avoided when designing training curricula (Thibault et. al, 1998).

1. Trying to teach too much. *Break the material down into understandable parts.*

2. Trying to teach too fast. *It is better to cover less material in depth; also, consider each person's learning speed.*

3. Lack of communication concerning training plans. *Have formal plans established.*

4. Failure to recognize individual differences. *Adjust accordingly.*

5. Failure to provide practice time.

6. Failure to show other employees the big picture.

7. Failure to give positive reinforcement. *Need to use both external and internal motivators.*

8. Intimidating employees.

9. Lack of common vocabulary. *What is being taught?*

10. The Pygmalion effect. *Expectations must be realistic and appropriate. Do not pre-judge; skewed perceptions might lead to erroneous evaluations.*

Training Methods

The training method that is selected will depend on:

- The people to be trained
- The type of facilities available
- Cost in relation to the funds allocated
- The basic philosophy of the instructional personnel
- The urgency of the situation

No ideal approach or single method exists. Some of the more commonly used methods are (More and Wegener, 1996, pp. 432–433):

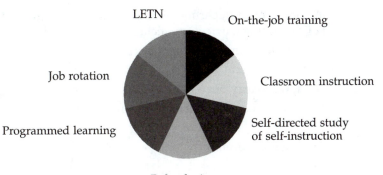

Common Training Methods

1. *On-the-job training:* has its good and bad sides. In less sophisticated departments, it is a trial-and-error method and can create problems.

2. *Classroom instruction:* easy, cost-effective, maximum number of trainees with a minimum number of instructors. Must be supplemented with:

   - Discussions
   - Demonstrations
   - Simulations
   - Role-playing

   Classroom instruction is more useful for conveying general information than it is for training. It is tailored primarily to the speed of presentation rather than the trainees' capacity to learn. In addition, the success of this method depends on factors like:

   - Complexity of the subject
   - Size of the class
   - Instructor's knowledge
   - Technical proficiency
   - Personality
   - Interpersonal skills
   - Overall teaching ability

3. *Self-directed study or self-instruction:* most effective for familiarizing the trainee with documents such as rules and regulations, as it is totally dependent on the motivation of the trainee. To acquire complex knowledge, most people need direction and interaction with "significant others."

4. *Role-playing:* interactive simulation; to simulate, as much as possible, those incidents that trainees are likely to face while on the job. Hands-on experience in a practical setting permits them to experience real-life effects.

In some police departments it is a mandatory method used in the basic police training curriculum, including, for example:

- How to confront a violent intruder
- How to conduct a body search
- How to carry out a vehicle stop

The key to success in this method includes:

- Problem identification
- Scenario planning
- On-site management
- Feedback

5. *Programmed learning (also called Programmed instruction or self-study method:* proven to be quite useful in transmitting information or skills. The instructor is a structured workbook or computer with a set of questions and answers that lead to feedback–skills development interactive programs: for example, "shoot/don't shoot" computerized range training or firearms training simulations.

6. *Job rotation:* important element in a comprehensive police personnel training. The trainees are assigned to each unit for a short period of time—for a tour of duty.

7. *Law Enforcement Television Network (LETN):* professionally produced videos of various training segments offered on a repetitive schedule on a 24-hour basis.

The British *Probationer Training Manual* (British Police Training Center, 1999) adds a lot of valuable information in regard to effective teaching methods that can be used in an academy setting to maximize the performance of recruits. Following is a summary of the methodologies identified by the British police as potential approaches to police training.

Teaching Methods in the British Police Academy

1. Student-centered methods
    - Facilitation
    - Experiential learning cycle
    - Adult learning cycle (a model shows four stages that a learner may experience: disorientation, exploration, reorientation, and equilibrium)
    - Giving and soliciting feedback
    - Briefing and debriefing
2. Trainer-centered methods
    - Presentation
    - Demonstration

Groups

Student-centered methods

Activities for groups

Trainer-centered methods

Group learning methods

Individual study

Teaching Methods: In the Academy

3. Individual study
   - Prereading information given to learners before a lesson or course.
   - Self-study.
4. Group learning methods
   - Knowledge checks.
   - Large-group discussions.
   - Round robins: a discussion method in which each member of a group is invited to comment in turn.
   - Free-boarding/board-blasting: a method of sharing ideas within a group. The trainer introduces a topic and the students give short responses which they feel are associated with the topic. These are noted on a pen board, then reviewed and discussed further. This approach, also known as *word-storm*, can be used as a problem-solving technique.
   - Group-forming exercises.
5. Activities for groups
   - *Case exercise:* a short set of circumstances related to the students. This might typically be a written scenario, video clip, or a news article. The students can use the case exercise to test their understanding of the law or to explore wider policing issues.
   - *Case study:* a detailed account of a policing situation based on a real incident. Case studies provide the views of various people involved in the incident to assist in considering the issues from more than one perspective. Case studies designed for use in police training; normally anonymous and based around a fictitious location.
   - *Paper feed:* case exercises where students are provided with information in progressive stages.
   - *Structured experiences.*
6. Types of groups
   - *Tutorials:* normally one-to-one.
   - *Dyads (or pairs):* students working in twos.

- *Buzz groups:* small numbers of students working together.
- *Pyramid groups:* learning activities in which students start working on a task in small groups, then combine into larger groups to work on the same or a similar task as the session progresses.
- *Horseshoe groups:* a seating arrangement that allows students to work together in groups while maintaining a view of the front of the classroom, and contact with the trainer.
- *Goldfish bowl:* a classroom activity in which some students participate in a task while others sit "outside" the activity and observe.
- *Crossover groups:* an exercise in which students form groups for one part of a task, then move to new groups for the next phase of the task; the second groups are arranged to ensure that they all contain at least one student from each of the first set of groups.
- *Syndicates:* set groups that reform regularly during a course.

Action Planning

Action planning consists of:

- Setting targets or objectives to accomplish some task or tasks
- Establishing standards of performance
- Establishing time scales that act as deadlines
- Allocating personal responsibility for these actions

An action plan is an agreed-upon, specific, and achievable plan of how personal and professional development can be addressed within a given time limit. It reflects two elements:

1. The plan
2. The action to achieve the objectives set out in the plan

An action plan should be "SMART" (British Police Training Center, 1999):

- Specific
- Measurable
- Achievable
- Realistic
- Timed

The learning process can be divided into three categories, which can facilitate learning of certain topics while keeping applied in conducive environments. The learning process is comprised of three stages (British Police Training Center, 1999):

1. *Cognitive:* learning results that emphasize intellectual outcomes such as
   - *Knowledge.* The first level of cognitive learning requires the learner to recall and recognize items such as terminology and facts.
   - *Understanding.* The second level addresses the learner's ability to comprehend the knowledge obtained to the extend that he or she can use it in limited situations.
   - *Thinking.* The third level is the application phase, where the learner begins to apply to the real world the knowledge learned.
2. *Psychomotor:* learning results that emphasize motor skills such as operating a two-way radio or typing the end-of-shift report. Several skills are involved in this learning process:
   - Perceive (covers a person's awareness through sensory perception)
   - Tell
   - Show
   - Practice
   - Evaluate
3. *Affective:* learning results that emphasize intrinsic feeling and emotions that a person has toward a particular subject, such as interests, attitudes, and methods of adjustment.

While devising a specific training module it is essential to incorporate the various learning stages into the planning process and to familiarize trainees with the way they absorb the materials conveyed to them.

THE ADULT LEARNER

We looked previously at general theories of learning, but models that focus primarily on adult learning have also been developed and deserve consideration and inclusion in the planning phase of training. A number of theories to consider are described in Table 3–5.

Course Content: Task Analysis

Roberg and Kuykendall (1997) emphasize the importance of the content of training programs. As much as understanding a learning process is crucial to devising an effective training module, its content can either be extremely helpful or completely detrimental to the way we perceive and learn. Training programs should be based on two common assumptions:

1. The mission statement of the department and on ethical considerations.
2. What an officer actually does every day.

Table 3–5 Adult Learning Theories

Theory	Developed by:	Main Concept
Individual autonomy theory	Knowles (1978)	In this theory the focus is on characteristics that have implications for both the learner and teacher. It is postulated that for learning to be effective, it is important that the adult learner be allowed to choose whether or not to seek the training.
Internal development	Havighurst (1948)	This model focuses on how internal development into adulthood and orientation to the outside world take place. Three stages are identified: expansion/extroversion of competence, a period of reorientation, and a change from active to passive mode.
Programmed values	Massey (1979)	This theory emphasizes the importance of having a facilitator who is aware of the different values and motivational factors of adults. It is proposed that most values are generally programmed into people by the age of 10.
Career anchors	Schein (1992)	Career anchors consists of a combination of competence, motivation, and values that a person *never* gives up. They represent the self. For best performance, these anchors have to complement the occupation being trained for.

The subject matter to be taught in the academy should be based on task analysis of the jobs to be performed by recruits. Today, it is still more an exception than the rule—with police training based more on legal requirements and experience.

Some departments have tried to identify police tasks that are important to the job and to base training on those tasks. The importance of such an approach cannot be overstated. Issues to be addressed include (Roberg and Kuykendall, 1997):

1. The degree to which reality is presented and discussed
2. The image of the police presented
3. The weight given to tradition (traditionally, training underrepresented order maintenance and social service aspects of police work and has overrepresented strict law enforcement activities)
4. The amount of time devoted to a particular subject, which emphasizes to recruits the importance attached to that subject by the department

The Integrated Curriculum

The concept of an integrated curriculum is important in education and training because it is generally recognized that its use can enhance the learning environment and the achievement of learning outcomes. A common misconception is for people to consider that the curriculum consists only of content: for example, just knowledge or a syllabus. Curriculum is not just knowledge but also the value system within which the knowledge is presented and the way in which presentation takes place.

Pring (1976a,b) identifies four distinct modes of integration with the curriculum:

1. *Integration in correlating distinct subject matter*. This situation exists where there is a dependence of one or more subjects on another. An example of this mode of integration in the context of police training is the teaching of law/legislation to provide trainees with a foundation of knowledge that informs later practice on the street, such as the use of discretion.

2. *Integration through themes, topics, and ideas*. This occurs where principles are identified that are inclusive enough to unite diverse subject matter, thus offering alternative ways of organizing knowledge. The primary over-reaching principle or set of principles that unite all facets of a trainee curriculum is the value and mission statement of a police organization. Use of this statement to unite diverse subject matter is a good example of this mode of integration.

3. *Integration in practical thinking*. This can be reflected in examples such as living in a multiracial society, the relations between the genders, making sense of violence, and living in a democracy, all of which require understanding and judgment as well as values. Recruit training and police training in general takes place against a background of all the factors in the example. The value judgments and understanding necessary to function effectively as a police officer in today's society will inform every facet of the curriculum that prepares police officers to carry out their roles.

4. *Integration in the learner's own interested enquiry*. In this case, the activities that a learner is likely to learn most from are those that arise from concerns that give shape to the learner's current view of the world, the problems the learner regards as important, and the questions the learner wants answered. The focus here would be on the recruits' needs in a learner-centered environment.

Each of the modes of integration covered here could be a facet of an integrated curriculum. They do not have to be present at the same time, but effective integration can enhance the learning environment and aid in the achievement of curriculum aims (British Police Training Center, 1999).

Program Philosophy and Instructional Methods

Problem solving and analytical thinking, so inherent in the daily life of a police officer, are, unfortunately, not the guiding principles in the program philosophy and instructional methods adapted by many police academies. However, with the ever-increasing recognition of the complexity of the police role and hence the necessity to utilize both training and education to prepare police recruits, the traditional distinction between the two approaches has to become less significant to achieve the desired outcome.

The instructional methods used in the academies play a very important role in the delivery and retention of the material presented and conveyed. The instructional methods are based on two schools of thought: behavioral and Gestalt. *Behavioral theorists* believe that learning is a function of stimulus and response (S-R), or trial and error. Desired behavior, or the learning of knowledge and skills, is rewarded; undesired behavior, or the failure to learn, is punished or ignored. Trainees are given points or some form of credit for desirable behavior, such as attendance, promptness, or performance. *Gestalt theorists* do not support the stimulus–response approach; they maintain that learning is more cognitive, involving gaining or changing insights, points of view, or thought patterns. More emphasis is placed on the analytical processes of reasoning and problem solving. The two theories can be differentiated in that behaviorist teachers want to change the behaviors of students in a significant way, whereas Gestalt-oriented teachers want to help students change their understanding of significant problems and situations.

Instructional methods are influenced by teaching philosophies, which in turn are influenced by learning assumptions. Two contrasting philosophies are pedagogy and andragogy. *Pedagogy* involves the one-way transfer of knowledge, usually by lecturing on facts, in which absolute solutions are expected to be memorized. *Andragogy* stresses analytical and conceptual skills and promotes the mutual involvement of instructor and student. Knowles (1970) describes pedagogy as the art and science of teaching children and andragogy as the art and science of helping adults learn. Andragogy is based on four assumptions about the characteristics of adult learners:

1. The learner's self-concept moves from that of a dependent personality toward that of a self-directing human being.
2. The learner accumulates a growing reservoir of experience, which becomes an increasing resource for learning.
3. The learner's readiness to learn becomes oriented increasingly toward the developmental tasks of his or her social roles.
4. The learner's time perspective changes from one of postponed application of knowledge to immediate application, and accordingly, his or her orientation toward learning shifts from subject-centeredness to problem-centeredness.

The four andragogical assumptions might have very significant implications on the design of a training program and environment:

1. The learning climate (psychological as well as social) should be cooperative and nonthreatening.
2. Students and teachers should view themselves as joint inquirers (nonstressful).
3. Teaching–learning transaction should be seen as a mutual responsibility of both the teacher and the learners.
4. A shift should occur away from transmittal techniques toward more participatory techniques.
5. The orientation to learning should be problem centered and focus on the practical concerns of the learners.

Andragogy incorporates a Gestalt orientation; pedagogy uses a behavioral approach. Since insight, analysis, and problem solving are important aspects of police training, academies should consider adapting both andragogical techniques and a pedagogical approach, an effective combination for a good part of their training program. Whereas adragogy could be very effective in development of the skills related to the community-oriented philosophy, pedagogy can be equally effective for activities that require memorization of laws and behavioral techniques such as traffic stops (Roberg and Kuykendall, 1997).

Effective Learning Environments

The final part of this chapter is devoted to effective learning environments, which deserve some consideration by the people responsible for the development of training modules. The main concepts are summarized in Table 3–6.

Both experiential learning camps produced questionnaires that enable learners to assess the category or categories of learning style that they exhibit at the time of assessing themselves. In the field of training, experiential learning as a technique can be very powerful. According to the British Police Training Center (1999):

1. Experiential learning creates opportunities for meaningful learning, involving understanding attitudes and overt behavior. Learners develop their own insights, interpretations, and conclusions after reflecting on their experience.
2. Learning is an action-based, dynamic, and participatory process.
3. Learning can be classroom-based or operationally situated, vicarious or otherwise.
4. Trainers are required to create opportunities for learning and then support the learners to make sense of the learning that has taken place.
5. Training is aimed at linking theory with practice to make teaching relevant to a workplace situation.
6. Trainers brief and debrief learners to ensure that the learning taking place is made explicit.

Table 3–6   Effective Adult Learning Models

Model	Developed by:	Main Concept
"Knowing-in-action"	Schon (1987)	The learning environment should be risk free but realistic, such that each person realistically can reflect on his or her own actions. The setting of learning is the practicum, in which seeing, thinking, and doing take place.
Action theory	Revons (1980)	The learning process can be viewed as a cycle consisting of planning, acting, observing, and reflecting, in which there has to be an evaluative process by the learner and every phase has to be completed for it to be effective. The process can start in any phase.
Personal knowledge	Polanyi (1962)	This theory emphasizes that knowledge is achieved by a person's unique character and the rational reflection of his or her own action.
Experiential learning cycle	Kolb (1984)	Kolb describes four learning styles that are based on experiential learning: 1. *Diverger:* emphasizes concrete experiences and reflective observation 2. *Assimilator:* combines reflective observation and abstract conceptualization 3. *Converger:* combines abstract conceptualization and active experimentation 4. *Accommodator:* combines active experimentation and concrete experiences

Once again the British Police Training Center has come up with a set of recommendations that should be taken into consideration by training directors and module developers in the United States. The following ideas summarize the major concepts and theories related to the learning environment.

1. Support should be given that helps learners link their current practice with the content of the learning activities.
2. Students' preferred learning styles should be identified and, when appropriate, learners should be encouraged to adopt a more balanced style.
3. A learning environment of mutual trust, respect, and agreed levels of confidentiality should be established.
4. Training in the performance of specific complex skills should include:
   - The rationale for their use
   - An account of their nature and purpose
   - Demonstrations of skills being employed, preferably in a job setting
   - On-the-job coaching, ideally by peers

5. Learners should be encouraged to develop a critical understanding of their work/job.

6. Performance on the job is an essential and major component of the learning experience.

7. The existing experience of learners should be acknowledged and opportunities should be provided for the knowledge of their practice to be shared.

8. The learners' experience should be challenged through the provision of opportunities to develop new insights and frameworks for guiding subsequent action in the job.

SUMMARY

A number of learning theories and philosophical themes were presented to illustrate the complex nature of adult learning. Advantages and disadvantages of training versus education were contrasted against the needs of today's police departments and police officers on the streets. Individual differences and deficiencies, as well as errors to avoid, were outlined. The multitude of concepts and approaches presented in this chapter could be slightly overwhelming at first glance given the fact that complex theories were presented in an abbreviated manner. However, a reader interested in a given approach is directed to the relevant source, identified by the concept defined, to familiarize himself or herself in depth with the specific area of interest. The main goal of the chapter was to provide the reader with a number of ideas and themes which can serve as a constructive guide to a more profound look at what needs to be taken into consideration while designing an effective training module. An insight into the world of adult learning can certainly provide for a fruitful discussion around a number of critical but equally important factors: the length, content, and methodology of the module.

REFERENCES

Bandura, A. (1977). *Social Learning Theory*. Englewood Cliffs, NJ: Prentice Hall.

British Police Training Center. (1999). *Probationer Training Manual*.

Cohen, R. G. (1972). Outreach and advocacy in the treatment of the aged. *Social Casework, 55,* 271–277.

Dilts, R., and Mayer-Anderson, M. (1980). *Neurolinguistic Programming in Education*. Cupertino, CA: Meta publications.

Festinger, L. (1957). *A Theory of Cognitive Dissonance*. Stanford, CA: Stanford University Press.

Griffiths, C. T., Whitelaw, B., and Parent, R. B. (1999). *Canadian Police Work*. Scarborough, Ontario, Canada: International Thomson.

Havighurst, R. J. (1948). *Developmental Tasks and Education*. Chicago: University of Chicago Press.

Heckel, R. V., Wiggins, S. L., and Salzberg, H. C. (1962). Conditioning against silences in group therapy. *Journal of Clinical Psychology, 18,* 216–217.

Herzberg, F. (1959). *The Motivation to Work*, 2nd ed. New York: Wiley.

Jacobs, R. T., and Furhman, B. S. (1984). *The Concept of Learning Style*. San Francisco: Jossey-Bass.

Knowles, M. S. (1970). *The Modern Practice of Adult Education: Andragogy versus Pedagogy.* New York: Association Press.

Knowles, M. S. (1978). *The Adult Learner: A Neglected Species,* 2nd ed. Houston, TX: Gulf Publishing.

Kolb, D. (1984). *Experiential Learning.* London: Prentice Hall.

Maslow, A. H. (1943). A theory on human motivation. *Psychological Review, 50,* 370–396.

Massey, M. (1979). *The People Puzzle: Understanding Yourself and Others.* Reston, VA: Reston Publishing.

Merrill, D. W., and Reid, R. H. (1981). *Personal Style and Effective Performance.* Radnor, PA: Chilton Book Co.

More, H. W., and Wegener, W. F. (1996). *Effective Police Supervision,* 2nd ed. Cincinnati, OH: Anderson Publishing.

Pavlov, I. P. (1927). *Conditioned Reflexes: An Investigation of the Physiological Activity of the Cerebral Cortex.* London: Oxford University Press.

Perls, F. S. (1969). *Gestalt Therapy Verbatim.* Lafayette, CA: Real People Press.

Plunkett, R. W. (1992). *Supervision: The Direction of People at Work.* Needham Heights, MA: Allyn & Bacon.

Polanyi, M. (1962). *Personal Knowledge: Towards a Post-Critical Philosophy.* London: Routledge & Kegan Paul.

Pring, R. (1976a). *Knowledge and Schooling.* Shepton Mallet, Somerset, England: Open Books.

Pring, R. (1976b). *The Integrated Curriculum, Unit 12 of Curriculum Design and Development.* Milton Keynes, Buckinghamshire, England: Open University Press.

Revons, R. W. (1980). *Action Learning.* London: Blond and Briggs.

Roberg, R. R., and Kuykendall, J. (1997). *Police Management,* 2nd ed. Los Angeles: Roxbury Publishing.

Rogers, C. R. (1951). *Client-Centered Therapy: Its Current Practice Implications, and Theory.* Boston: Houghton Mifflin.

Rotter, J. B. (1966). Generalized expectancies for internal versus external control of reinforcement. *Psychological Monographs: General and Applied, 80,* 1–28.

Schein, E. H. (1992). Career anchors and job/role planning: the links between career planning and career development. In D. H. Montross and C. J. Shinkman (eds.), *Career Development: Theory and Practice.* Springfield, IL: Charles C Thomas.

Schon, D. A. (1987). *Educating the Reflective Practitioner: Toward a New Design for Teaching and Learning in the Professions.* San Francisco: Jossey-Bass.

Skemp, R. R. (1962). The need for a schematic learning theory. *British Journal of Educational Psychology, 32,* 133–142.

Skinner, B. F. (1965). *Science and Human Behavior.* New York: Free Press.

Sommer, R. (1967). *Personal Space: The Behavioral Basis of Design.* Upper Saddle River, NJ: Prentice Hall.

Taylor, M. (1979). Adult learning in an emergent learning group: toward a theory of learning from the learner's perspective. Ph.D. dissertation. University of Toronto.

Thibault, E., Lynch, L. M., and McBride, R. B. (1998). *Proactive Police Management.* Upper Saddle River, NJ: Prentice Hall.

Timm, H., and Christian, K. E. (1991). *Introduction to Private Security.* Pacific Grove, CA: Brooks/Cole.

Tolman, E. C. (1932). *Purposive Behavior in Animals and Men.* New York: Century.

Tolman, E. C. (1951). *Collected Papers in Psychology.* Berkeley, CA: University of California Press.

Tuckman, B. W., and Jensen, M. A. (1977). Stages of small-group development revisited. *Group and Organization Studies, 2,* 419–427.

Watson, J. (1930). *Behaviorism,* rev. ed. New York: W.W. Norton.

Whisenand, P. M., and Rush, G. E. (1993). *Supervising Police Personnel,* 2nd ed. Upper Saddle River, NJ: Prentice Hall.

Zajonc, R. B. (1965). Social facilitation. *Science, 149,* 269–274.

ACADEMY TRAINING

Most police cadets complete a minimum of only 400 hours of basic training before becoming a sworn officer. An exceptionally well-trained cadet will receive only about 800 hours. Attorneys, on the other hand, receive over 9,000 hours of instruction and doctors over 11,000 hours. Even embalmers, with 5,000 hours, and barbers, with 4,000 hours, receive more training than police officers . . . . The concern over the inadequacy of police basic training programs has resulted in one police scholar observing, "Doctors bury their mistakes, while lawyers send theirs to jail." Unfortunately, untrained police officers do a little of both. (Edwards, 1993, p. 23)

INTRODUCTION

This chapter provides a short overview of the basic academy training conducted in a number of chosen agencies. The goal of this presentation is to emphasize the problems associated with the ways a basic academy is conducted with regard to both content and length of the training. The three types of basic training—in-house, state, and regional—are presented through a closer look at the instructional modules offered by each academy. As stated at the onset of the book, it is quite probable that both the length and content of any given example presented in this chapter are no longer valid. Nevertheless, the agencies represent the trends in police training during the last years of the twentieth century and the first steps into the new millennium.

Any specific allocation of hours or topics illustrates themes that are deemed important enough to be included in the minimum hours allowable for training. As Edwards (1993) insightfully points to the deficiencies in hours of training allocated to police academies, the goal of this chapter is to point to another dimension of deficiency—the content of the modules and their relevance to today's policing.

APPROACHES TO ACADEMY TRAINING

Today, the term *police academy* usually refers to three main types of police academies in the United States: agency, regional, and college-sponsored. *Agency schools* are generally found in large municipal areas or are established for the state police or highway patrol. *Regional academies* handle the training functions for both large and small departments located in a designated geographical area. *College-sponsored training academies*, which operate on the premises of postsecondary institutions, particularly community colleges, allow a person to take police training and earn college credit (Thibault et al., 1998).

Modern police training has come a long way since the times when the statement "Ignorance of police duties is no handicap to a successful career as a policeman" (Reith, 1975) mirrored the grim state of the police profession. Fosdick (1969, pp. 298–299) described some examples of police training in police departments that he visited during the second decade of the twentieth century: "Probably the most ambitious police school at the present time is in Berkley, California. Here, the class, which meets one hour a day, takes three years to complete the courses included in the curriculum. The New York school involves two months of full-time instruction; in Chicago, Philadelphia, St. Louis, Detroit and Newark, the training period is four weeks; in Cleveland it is three. in Cincinnati and Louisville only part of the day is devoted to school work, the remainder being spent in the performance of regular duty."

The Berkley police school owed its ambitious program to its police chief, August Vollmer. Appointed to the position in 1909, he held that office continuously until his retirement in 1932. In 1914 he set up the Berkley Junior Police and shortly thereafter reorganized the Berkley Police School curriculum, a training program which over the next several decades was copied in whole or in large part by police agencies not only in many U.S. states but also by a number of police departments abroad (Stead, 1977). However, despite the efforts to define policing as a profession in most cities, the process of reform was painfully slow. Some cities did not offer any meaningful training until the 1950s (Walker, 1999). The situation in smaller jurisdictions was even more discouraging. In New Jersey, for example, in 1961 the Sussex County Police School offered a basic training of 2 hours per night "as long as needed," the Union County Police School offered one week of school, the Burlington County Bridge Commission Police School offered a 10-hour course, and Morris and Essex counties offered 2 hours per night for eight weeks (Matheny, 1996).

In 1959 in California a training council first established police officers' standards and training guidelines. Although this was a state rather than a federal initiative, within approximately four years, many states thoughout the country had developed their own standards for training (Christian and Edwards, 1985).

In 1967, the President's Commission on Law Enforcement and Administration of Justice recommended that a Peace Officer Standards and Training (POST) commission be established in every state. The POST commissions or boards were empowered to set mandatory minimum requirements and were funded appropriately so that they might provide financial aid to governmental units for the implementation of established standards. The two important charges of the POST commissions were (Bennett and Hess, 1996):

1. Establishing mandatory minimum training standards (at both the recruit and in-service levels) with the authority to determine and approve curricula, identify required preparation for instructors, and approve facilities acceptable for police training
2. Certifying police officers who acquired various levels of education, training, and experience necessary to adequately perform the duties of the police service

McManus (1970) surveyed 360 agencies, and the Chicago Police Department devoted 27- to 40-hour weeks to training (having the longest period of training of the agencies surveyed) and Kalamazoo, Michigan had the shortest, with 120 hours or three weeks of training. Over the past 30 years, significant changes have been made in preservice training. In 1993, the Bureau of Justice Statistics surveyed more than 12,000 municipal and county law enforcement agencies. They found that on average, departments required 640 training hours of new officer recruits, including 425 classroom training hours and 215 field training hours (Edwards, 1993, p. 5). Despite the recent findings, caution must be used prior to generalization about the nature of academy training at the end of the twentieth century. The key word being "average" does not take into consideration the implications of the less than average number of hours after which a given recruit is sent to perform as a police officer. The main objective of this chapter is to portray a general picture of the number of weeks and hours devoted to training in a given jurisdiction and a basic outline of topics, as well as to analyze the content of curriculums gathered from a number of academies throughout the country. Of particular interest is the Washington, DC agency training academy, the only academy in the United States that is not subjected to POST standards. Frost and Seng (1984) replicated an earlier study and surveyed training curriculums in 33 police departments of various sizes around the country. They discovered that the entry-level training programs grew from an average of 342 hours in 1952 to an average of 633 hours in 1982. However, much of the content had not changed. The percentage of time devoted to military drill went down but was replaced by a rise in the time devoted to general physical fitness, while the time devoted to firearms training remained about the same. Today, newer topics have been added and are discussed later. Nationally, most training programs can be categorized in the following areas (Thibault et al., 1998):

1. *Administrative procedures:* includes quizzes, graduation, and instruction on note taking
2. *Administration of justice:* includes history of law enforcement, police organization, probation, parole, and social services
3. *Basic law:* includes constitutional law, offenses, criminal procedure, vehicle and traffic law, juvenile law and procedures, and civil liability
4. *Police procedures:* includes patrol observation; crimes in progress; field notes; intoxication; mental illness; disorderly conduct; domestic violence; police communication; alcoholic beverage control; civil disorder; crowd and riot control; normal duties related to traffic enforcement, including accidents and emergency vehicle operations; criminal investigations, including interviews and interrogations; control of evidence; and various kinds of cases, such as burglary, robbery, injury, sex crime, drugs, organized crime, arson, and gambling
5. *Police proficiency:* includes normal firearm training, arrest techniques, emergency aid, courtroom testimony and demeanor, and bomb threats and bombs
6. *Community relations:* includes psychology for police, minority groups, news media relationships, telephone courtesy, identification of community resources, victim/witness services, crime prevention, officer stress awareness, law enforcement family, and police ethics

Training programs should be based on specific goals. Common sense advises us that those goals should first and foremost promote the idea of providing recruits with the basic skills necessary to perform the functions of a police officer and to socialize him or her into the police profession. No police professional would argue with these basic concepts; however, the authors argue that the first and most important goal of academy training would be to convey to the "new believer" an understanding of what police work is all about. "Police work is a tainted occupation" (Bittner, 1980)—it is all about the use and misuse of coercive force. According to Bittner, the role of police is best understood as being a mechanism for the distribution of nonnegotiable coercive force employed in accordance with the dictates of an intuitive grasp of situational exigencies (1980, p. 28). At the other end of the spectrum, the philosophers and followers of community-oriented policing argue that police work is about providing services to the community and solving crime problems together with the community (the definition of COP offered here is, of course, simplistic, but it conveys the basic premise and concept of this philosophy). The introductory (and some more advanced) books in the field of police science define the role of the police in the society in an even more enigmatic manner. According to Wroblewski and Hess (1997:62), five basic responsibilities of most police officers are to:

1. Enforce laws
2. Preserve the peace
3. Prevent crimes

4. Protect civil rights and civil liberties

5. Provide services

In developing these themes further, Wroblewski and Hess offer a more insightful view into the basic concepts. Police officers are responsible for enforcing laws and for assisting in the prosecution of offenders; they are responsible for attempting to prevent a crime, protecting a citizen's constitutional rights, and are also often responsible for providing services (1997, pp. 63–67).

The three different approaches to the role of police in a modern society present a formidable challenge for the decision process that must go into formulating and defining the appropriate basic training curriculum that will at least afford recruits the basic and most desirable tools with which to begin a law enforcement career. The distinction between the *traditional skills* of policing required from the police officer before the community-oriented policing era (BCOP) and the *new skills* introduced by the community-oriented policing (ACOP) philosophy are emphasized through an analysis of training curricula and specific modules. The traditional skills are those related to the use of force, physical fitness, driving skills, self-defense tactics, and the like—in other words, skills related to physical agility more than to any other human skills. The new skills are those related to integrity, leadership, ethics, communication, cultural awareness, community policing philosophy, victim assistance, domestic violence, and others.

Prior to a discussion of the recommended training curriculum, including modifications of existing modules and formats, it is necessary to look at some of the current trends in agencies varying in location, size, and available resources. Additional training curricula not discussed in this chapter can be downloaded from the Internet and serve as class exercises.

Some of the agencies and their academies are described at length, whereas some are referred to only in general terms, and a number of specific topics are highlighted and discussed. This uneven representation of information itself is indicative of the state of police training. Although some academies were extremely open and cooperative in providing the author with the information requested, some denied access to material or made only general references to the topics and issues covered. As much relevant information as it was possible to gather is provided as well as other pertinent information. In-house academies usually offer a wealth of relevant information, including the demographics of the area. State and regional academies do not offer demographics, as they train officers from various areas of the state.

AGENCY A: WASHINGTON, DC, POLICE DEPARTMENT (IN-HOUSE ACADEMY)

The Metropolitan Police of Washington, DC have their own in-house training academy (but no POST standards). The Recruit Officer Training and Certification Program (ROTCP) is 26 weeks long, including two weeks on the street. It is intended to present fundamentals of modern police work and general and specific knowledge necessary to make the recruit a competent professional performer at the level of patrol officer. It is geared to assist in developing superior personnel, necessary to provide the future leadership of the department. To enhance leadership and team-building skills, a resident training component is included in the program beginning in the second week. The entire training program consists of lectures, workshops, visual aids, operations, review of course material, skills labs, self-directed learning, and field performance tasks. Emphasis is placed on performance-based training and community empowerment policing. Over 80 subjects make up the 16 training levels of the curriculum.

Program Goal. This program is a comprehensive and uniform method for training, supervising, guiding, and evaluating newly appointed members of the department. Recruits should leave this program with the knowledge and skills necessary to become patrol officers.

Training subjects

- Orientation and personnel issues — 40 hours
- *Level 1:* Organization of the Department — 49 hours
- *Level 1–2:* DC Code, Part I — 44 hours
- *Level 1–3:* DC Code, Part II — 44 hours
- *Level 4:* Criminal Procedure, Part I — 41 hours
- *Level 5:* Criminal Procedure, Part II — 59 hours
- *Level 6:* Investigative Patrol Techniques — 58 hours
- *Level 7:* Handling Property — 25 hours
- *Level 8:* Unique Patrol Situations — 26 hours
- *Level 9:* Traffic Regulations — 18 hours
- *Level 10:* Traffic Enforcement — 36 hours
- *Level 11:* Use of Force: Firearms — 68 hours
- *Level 12:* Vehicle Skills Training — 42 hours
- *Level 13:* First Responder — 33 hours
- *Level 14:* Use of Force: Defensive Tactics — 85 hours
- *Level 15:* Behavioral Sciences — 37 hours
- *Level 16:* Civil Disturbance — 35 hours
- *Level 17:* Physical Training — 52 hours

The use of force, levels 11 and 14 (all together 153 hours) indicates that the primary role of the police is closer to Bittner's definition than to any other, including the MPD's own goals. Level 17, physical training (52 hours) seems to point to that direction as well.

Level 15: Behavioral Sciences

- *Lesson objective:* To enable recruits to gain knowledge of and have an opportunity to apply the fundamentals, principles, and techniques recommended when restraining individuals.

- *Performance objective:* To accomplish the following performance objectives:

Task	Performance/Description	Hours
1	Diversity awareness/sensitivity training	8
2	Bias-related crimes	3
3	Victim assistance	3
4	Community-empowered policing	3
5	Domestic violence	20

A question remains open: Is diversity training really accomplished in 8 hours of instruction? Firearms training can obviously be accomplished in 68 hours of instruction—otherwise, recruits would not be trusted with life and death devices. But do they really meet performance objectives in the victim assistance module after 3 hours of lecture?

AGENCY B: CHARLOTTE–MECKLENBURG, NORTH CAROLINA, POLICE DEPARTMENT (IN-HOUSE ACADEMY)

In 1971, the General Assembly established the North Carolina Criminal Justice Training and Standards Council. By 1973 the council had mandated a minimum basic training requirement of 160 hours to be completed by officers within their first year of employment. On October 1, 1978, this standard was increased to 240 hours. The standardized curriculum applicable to the 240 hours was designed by a cooperative venture of the Basic Law Enforcement Training Consortium. The consortium consisted of representatives of the Department of Community Colleges, the Institute of Government, the Training and Standards Division, the North Carolina Law Enforcement Training Officers' Association, and the North Carolina Justice Academy. In 1979, the council was reconstituted as the Criminal Justice Education and Training Standards Commission. The commission expressed its concern over the 240-hour mandated curriculum, which was subsequently analyzed and a new basic training curriculum developed in 1983. In 1984, the newly formed North Carolina Sheriffs' Education and Training Standards Commission adopted its training requirements, with additional topics mandated for deputies. In 1986, the Training Standards Commission approved the formation of a basic Training Revision Committee to monitor the course and recommend revisions as necessary. The committee, which replaced the consortium, is comprised of administrators, trainers, and practitioners in the law enforcement field. The committee meets regularly to consider relevant and important issues that may affect this curriculum. The Basic Law Enforcement Training Curriculum, on which the CMPD Academy's training is based, was last revised in 1992.

The school director may exceed the minimum hours mandated by the state. He or she can devise rules, regulations, or policies consistent with applicable laws, which are in fact more stringent than those contained in the original program. The school director or his or her designated representatives has to point out any variations between state-required minimum standards and additional or more stringent standards utilized by the accredited institution or agency that is providing the basic training program. There are a minimum of 31 topical areas, including orientation, in the basic training course, and trainees are expected to become proficient in each topic. Integrity-related topics, despite being highlighted in each training module, are emphasized specifically in the following modules, in sequence of instruction:

Module 1   Course Orientation
Module 2   Constitutional Law
Module 5   Law Enforcement Communications and Information Systems
Module 9   First Responder
Module 18   Special Populations
Module 30   Dealing with Victims and the Public
Module 31   Ethics for Professional Law Enforcement

Module 31 consists of 4 hours of instruction whose purpose it is to address and introduce principles of professional and ethical conduct in the law enforcement community. The objectives to be achieved include defining ethics and morals, professionalism, and duty, with explanations of the basis for the Law Enforcement Code of Ethics, the Canons of Police Ethics, and the Oath of Office. Among the topics discussed is that of gratuities. Students are presented with a scenario and discuss the differences between a gratuity and a bribe, the ethical aspects of the differences, and the part played by the officer's involvement. The concepts of loyalty and personal integrity, honesty and truthfulness, and use of force and use of deadly force are discussed and analyzed based on scenarios and videotapes. Students are given the opportunity to ask additional questions and to explore issues not raised or discussed during the module.

The CMPD expanded on the basic state-mandated training curriculum by both adding a significant number of subjects and redesigning the modules. The Basic Law Enforcement Training Curriculum is divided into 77 subject modules. Of special interest in this chapter are the following modules:

- Breaking the Barriers (introduced during the seventeenth week of instruction for the duration of 4 hours)
- Child Development—Community Policing (also during the seventeenth week for 2 hours)

- Community Awareness Training (during the seventeenth week for 4 hours)
- Community Policing—Introduction (during the first week for 2 hours)
- Community Policing (during the tenth and fourteenth weeks for 5 hours)
- Community Policing—Problem Solving (during the eighteenth week for 16 hours)
- Dealing with Victims and the Public (during the second week for 4 hours)
- Domestic Violence Training (during the seventeenth week for 4 hours)
- Ethics for Professional Law Enforcement (during the sixteenth week for 4 hours)
- Special Populations (during the first week for 12 hours)
- Victim Assistance (during the fourth week for 1 hour)

Curriculum-Based In-Service Training. The CMPD began to deliver its in-service training under the curriculum-based training (CBT) format in January 1997. The CBT system resulted from the training academy's desire to meet three goals and objectives:

1. To develop an in-service training system promoting the department's mission
2. To create an ongoing training curriculum enhancing the employees' personal and professional development
3. To deliver education and training that strengthens the employees' skills, knowledge, and abilities

The yearly requirement for all sworn officers in the Charlotte–Mecklenburg department is 40 hours, consisting of 24 hours of mandatory training and 16 hours of electives. Employees receive three types of training in CBT:

1. *Departmental training:* any training provided to employees by the CMPD
2. *Specialized training:* any training provided by outside providers
3. *Senior Police Officer Courses:* a separate voluntary educational incentive program

Under the CBT, officers can apply any specialized training they received toward elective credit. Specialized training counts as elective course credit during the calendar year taken. For example, officers assigned to special units receiving monthly training count that training as an elective course. Several units meet this requirement:

- SWAT team monthly training
- Bomb squad monthly training

- Canine unit monthly training
- DARE officers yearly training

In 1997, 18 sessions of "Service Through Accountability and Respect" (STAR) were offered, comprising a total of 144 hours of training. This class is mandatory for nonsworn employees and optional for sworn employees. Following is a sample of elective courses (related to integrity issues) offered and delivered in 1997:

- Problem Solving
- Dynamic Problem Solving and Creativity
- Advanced Problem Solving
- Elder Abuse

Following is a sample of mandatory courses (related to the integrity issues) delivered in 1997:

- Introduction to Cross-Cultural Communication
- Customer Service

The CMPD definitely has a very impressive approach to its basic training, both in the topics included in the basic curriculum and the number of hours allocated to the topics. On the other hand, if one looks at the number of hours allocated modules related to the use of force, the math becomes embarrassing, especially when dealing with such a progressive and community-oriented academy.

- Defensive Tactics   24 hours
- Use of Force   4 hours
- Civil Disorder   12 hours
- Combat Shooting   7 hours
- Firearms   60 hours

This author would argue vehemently with any critic who concludes that firearms training requires 60 hours of instruction but ethics can be covered in 4 hours!

Agency C: St. Petersburg, Florida, Police Department (Regional Academy)

The St. Petersburg department sends its new recruits to a regional training academy, located in St. Petersburg, the St. Petersburg Junior College Criminal Justice Institute. During our research the current curriculum was the one revised in July 1997, the previous revision occurring in 1995. The Florida Criminal Justice Standards and Training Commission has the responsibility of developing effective and efficient job-related training programs for basic recruits in law enforcement, corrections, and correctional probation. To this end, a systems approach to curriculum development was adopted, which includes:

- An occupational survey to determine what tasks entry-level officers perform
- A task analysis to determine what an officer must do to perform each task
- An instructional analysis to determine what an officer must know to perform each task
- Learning goals and objectives to outline each competency topic
- Field testing of the learning goals and objectives to determine validity
- Collection of input for continual updating of curricula

This process resulted in revised basic recruit curricula, based on the job these officers perform. Revisions have been made in 1990, 1991, 1993, 1994, and 1995. Based on the most recent information, it was revised again in 1997, and finally, in 1999, therefore presenting a model case of a police training curriculum updated on a relatively frequent basis, a pattern to be commended. Although a regional academy is less under the influence of a given chief, this one, since it is located in the city of St. Petersburg, definitely affords the chief some possibilities of interference or at least input. From conversations with the staff of the academy it was apparent that such input indeed occurs from time to time. The Basic Law Enforcement Training Curriculum is divided into 12 courses, each divided into separate modules addressing a different topic. Of special interest to this chapter are the following modules (offered under the course title "Criminal Justice Legal 1"):

- Ethical and Professional Behavior
- Bribery
- Perjury
- Use of Force

The behavior module addresses, among other topics, issues related to dishonesty, brutality, prejudice, offering or accepting gratuities, violation of civil rights, and discourteous conduct. In addition, the force modules address moral and legal limitations on the use of force, suspects' demeanor and actions, and potential danger to officers and others, among other topics. Despite the fact that the topics addressed are of major importance to the issue of police integrity and police community relations, the number of hours devoted to training of these particular modules is minimal, on average, between 1 and 2 hours. (As noted above, the curriculum was revised in 1999, but the topics and hours presented here are from 1997.) As a comparison, 56 hours are allocated to the module devoted to weapons and chemical agent use, 66 hours to defensive tactics, and 43 hours to vehicle operation skills. It is true that among the modules one can find a number of hours devoted, here and there, to human relation skills, treatment of disabled populations, or mentally retarded persons. However, even when combined (and not scattered around different modules), the total number of hours devoted to the use of force and other restraint-related techniques, such as crowd control, far exceeds the number of hours that should be devoted to human relations skills, ethics, integrity, communication, and other skills related to the new role of police work, one in which the police becomes a service agency, serving and protecting the public.

The instructors at the training center are selected based on their experience and education, and include in-service officers, retired officers, and civilians.

AGENCY D: CHARLESTON, SOUTH CAROLINA, POLICE DEPARTMENT (STATE ACADEMY)

The city of Charleston, South Carolina Police Department is a relatively large police organization of about 320 sworn police officers. Based on the LEMAS Report (1996), only 0.7 percent of local police departments in the United States have a total number of sworn personnel ranging from 250 to 499 officers. The size of the department, and the nature of basic police training in South Carolina, offers only one alternative—the state academy.

The state of South Carolina offers mandatory police training at its state academy, which had a training program that lasted for about eight weeks and was recently revised and extended to nine and a half weeks. Having to send your new recruits to a state academy does not afford the chief any discretion regarding the mandatory curriculum. Such a situation, which is probably the worst of the three possible choices (in-house, regional, and state) mandates that very careful attention be placed on FTO training as well as other components of the in-service training. Despite the fact that the new recruits in Charleston are selected carefully, the length of academy training is barely adequate to master even the most general concepts of the theory of police work (an average length of academy training in the United States is almost double what is being offered by the state of South Carolina).

The topics included in the state academy's curriculum are related to critical issues of police work in either an elusive or an insufficient manner. For example, in the old training curriculum 8 hours were devoted to the topic of human relations and 20 hours to an enigmatic module entitled "Practical Problems." In the newly revised curriculum, the use of force is addressed for 2 hours and ethics for 3 hours.

AGENCY E: INDIANAPOLIS, INDIANA, POLICE DEPARTMENT (IN-HOUSE ACADEMY)

The Indianapolis department has 995 sworn officers and 265 civilian employees. The department has the responsibility for enforcing the law and keeping the peace in the 402.84-square-mile city with a population of 812,800. The in-house academy has 16 sworn and civilian instructors, all certified by the state of Indiana. Certified instructors have prior experience in teaching and training recruits.

To ensure that recruits concentrate on the training, the IPD prohibits recruits from holding outside employment during their training period. In addition, recruits are strictly prohibited from performing any direct law enforcement–related function requiring police power.

The training program is a mix of educational and physical regimens. Recruits must receive a minimum of 752 hours of training during a minimum 20-week period. All essential equipment is issued during the training period. The academy schedule is primarily Monday through Friday, 8:00 A.M. to 5:00 P.M., with some weekend and evening training required.

The courses and additional requirements and rules help recruits to become professional police officers. However, a closer look at the program reveals a number of problems. First, the training does not include a police ethics course, which is crucial in the prevention of police misconduct. Second, despite the fact that the department in its mission statement claims to be community oriented, its training curriculum does not include courses on community policing. A training module entitled "Human Behavior" has 76 training hours; in contrast, the use-of-force module has 157 hours and physical conditioning has 41 hours.

The IPD makes a commitment to community policing, but its basic training program lacks emphasis in that area. If community policing indeed is the primary orientation of the police force, training must begin in the academy.

Agency F: Suffolk County, New York, Police Department (In-House Academy)

The Suffolk County Police Department serves communities within the borders of Suffolk County. The force is composed of 2759 sworn officers and 603 civilians. Suffolk County's population is 1,356,896 residents, with a density of 1450 per square mile. In its mission statement the department stresses its obligation to provide professional services in the communities and to protect all persons and property in accordance with legal, moral, and ethical standards.

Police recruit training is conducted at the Suffolk County Police Academy, which is located in Suffolk County and run by the Police Department. The basic requirements for police officers in New York State are published by the New York State Bureau of Municipal Police. These are divided into eight parts, with a total of 52 topics. Each topic requires a specific number of minimum training hours. The minimum number of hours required is 510. That equals approximately 13 weeks, based on a 40-hour workweek. The Suffolk Police Academy's basic police course is 25 weeks long. For the first 15 weeks the emphasis is placed on discipline, physical training, defensive tactics, and classroom instruction on police procedures, law, investigations, and so on. The last 10 weeks of the program involve practical skills such as firearms qualification, emergency vehicle operation, officer safety, and emergency medical training certification.

Detailed curriculum from the academy was not available; however, based on the statement made by the commanding officer of the academy that they add about 30 percent to the hours required by the state, a simple calculation was performed. Part VI of the mandated (by the state) training module, community relations, requires 30 hours of training. Following are the required topics and hours allocated to training:

- Community Resources—Victim/Witness Services   3 hours
- Crime Prevention and Crime against the Elderly   4 hours
- Ethical Awareness   12 hours
- Cultural Diversity/Bias-Related Incidents/
     Sexual Harassment   5 hours
- Law Enforcement/Community Relations   2 hours
- Contemporary Police Problems   4 hours

If we add 30 percent to the above, we have 39 hours of training devoted to community relations, whereas the modules devoted to firearms training total 52 hours; defensive tactics, about 40 hours; and physical fitness, about 90 hours.

AGENCY G: NORTHERN VIRGINIA POLICE DEPARTMENTS (REGIONAL ACADEMY)

The Northern Virginia State Academy trains recruits from a number of police departments in Virginia. The most recent curriculum available was revised in 1999—offering, at least in theory, some hope for positive changes in the training orientation. Their basic law enforcement school curriculum consists of seven training modules that provide a total of 768 hours of training:

1. Course Administration   56 hours
2. Criminal Investigation   49 hours
3. Legal   138 hours
4. Patrol   116 hours
5. Skills   301 hours
6. Traffic   50 hours
7. Practical Exercises   58 hours

Looking for the community-related orientation, the author was able to identify the following topics and hours that seemed related to the new skills and orientation that police officers of the twenty-first century should acquire.

- Communicating with Developmentally Disabled Persons   1 hour
- Communicating with Hearing Impaired   3 hours
- Community-Oriented Policing   8 hours

- Community Relations   1 hour
- Domestic Violence   6 hours
- Diplomatic Immunity   2 hours
- Ethics and Discretion   4 hours
- Integrity   2 hours
- Interpersonal Communications   2 hours
- Intracultural Communications   4 hours
- Landlord/Tenant Disputes   2 hours
- Leadership   2 hours
- Mental Illness Cases   3 hours
- Telephone Communications   1 hour
- Victim–Witness Assistance Program   2 hours

For a total of 43 hours (plus some additional hours not specified devoted to practical exercises) the recruit is introduced to the new community-oriented skills. Although some of the modules are accompanied by practical exercises, some are not. For example, the author is curious to find out how much of leadership skills training a recruit can absorb in 2 hours of instruction, or how he or she can improve community relations after a 1-hour lecture. On the other side of the spectrum, however, defensive skills are acquired in 70 hours of instruction, coupled with 8 hours of use of force, 44 hours of firearms, and 70 hours of physical training. A total of 192 hours is devoted to the traditional skills, almost five times as much as to the new skills.

AGENCY H: THE IDEAL MODEL

This entire chapter could be filled with examples from numerous police training facilities around the country to prove the author's point. In the interest of time and effectiveness, however, the reader is asked to accept the notion that the examples provided in this chapter represent the norm in police training today rather than the exception. If the ideal police academy indeed exists somewhere in the United States and is not mentioned here, it will serve as another example of the critical issues in police training. The reason for this will be the lack of awareness on the part of the people involved in police training. Not one person the author and her research assistants spoke to mentioned such a model establishment. Therefore, the natural conclusion follows—the ideal establishment either does not exist or is very well hidden.

What follows is a model for a police academy, which is based on a generic formula that can be adopted by a state, regional, or in-house academy. The themes emphasized are not too farfetched or unrealistic; they simply point to what is missing, has to be added, has to be modified, and perhaps ignored. The realistic approach to the resources available will not allow full development of the ideal academy but given the number of hours each academy devotes to police training today, things can be changed, modified, and customized for the new philosophy of policing. All that is required is a shift in conceptual thinking and a true commitment to community-oriented policing—if the commitment is sincere.

In this chapter we provide only a basic contour for training modules, which in Chapter 15 is integrated with ideal FTO training, training methodologies, trainers, and learning environments. A number of key points need to be considered prior to devising any police training:

1. *What are the mandatory topics that need to be included in the curriculum, and what are the number of hours allocated to the topic?* The topics need to be evaluated from the standpoint of effective instruction that can be delivered in the number of hours allocated. For example, diversity training is mandated by POST as an 8-hour training module. The question to be answered is: Is it enough? The answer is: *No!* A wonderful textbook by Shusta et al. (1996), *Multicultural Law Enforcement*, offers a wealth of information on the topic. If an academy cannot afford to add additional hours to diversity training, the solution is to purchase this textbook and incorporate it in the 8 hours of training. Incorporate it such that in the final written examination the recruits will be forced to read, analyze, and memorize the important parts of this book. Alternatively, expand the number of hours devoted to the topic by reducing the number of hours devoted to another module: For example, subtract the additional 8 hours from the 60 hours devoted to physical fitness. The recruit will be as fit after 52 hours of training as after 60, but he or she will be much more aware of diversity issues after 16 hours of training rather than 8.

2. *How are the training modules sequenced?* In most cases training is offered in sequential modules (except for the Bergen County Police Academy). For example, once the topic is introduced, it is covered during the number of hours allocated to the module and never returned to. Even when a given topic is allocated the most ridiculous number of hours (e.g., leadership skills, offered for 2 hours only), it will be much more effective to break it into two segments and introduce it a few weeks apart, to enable the recruit to absorb, analyze, and return to the concept at least one more time. The law of "use and disuse" points to the fact that materials that are presented only once and never returned to will not be absorbed properly by the student. Other modules, which are offered during a number of hours, like the aforementioned diversity training, even with the minimal number of hours (8) can be broken down to either four or even eight segments, allowing a recruit to absorb the material in a much more profound manner. The problem of instructors who might not be available to deliver diversity training in eight separate segments, as most diversity trainers are brought from outside the police environment, can be resolved by training some in-house personnel. This particular topic can be taught by practically every trainer, as we represent such diversity in this country that each trainer can offer his or her perspective—all it takes are some pointers from an expert in the field and a good textbook.

3. *What and how much is added?* Most in-house, regional, and state academies do add to the POST requirements, some as little 10 percent, some double and even triple the basic requirements. When deciding what topics to choose, add, and supplement, a number of factors have to be considered:

   - Is the addition a post-hoc requirement? For example, a number of police departments institute a post-hoc diversity training after a major scandal erupts (e.g., in New York after the Diallo case).

   - Is the addition an outcome of a newly adopted philosophy? (This was the case with CAPS in Chicago.)

   - Is the addition an outcome of a personal belief of a newly appointed chief? (In the Charlotte–Mecklenburg department, Chief Nowicki, an ardent supporter of the COP philosophy, insisted on revision of the in-house training modules based on his orientation.)

   - Is the addition an outcome of civil litigation against the department? (The San Jose FTO program is an example.)

   After answering each question, the importance of a given module could be fully evaluated and the resources devoted to the module evaluated and allocated in a more appropriate manner. For example, if the addition to the POST requirements is based on the new philosophy of community-oriented policing, of the 100 hours added to the basic requirements, a very substantial number of hours has to be devoted to development of new skills and the number of hours allocated to traditional skills has to be balanced against that. Again, a number of hours removed from physical fitness can do miracles when converted into a leadership- or ethical/integrity-related module. The Chicago department almost doubled the state requirements when it introduced its revised training curriculum based on the CAPS (Chicago Alternative Police Strategy) philosophy. Unfortunately, only 29 hours of training is offered explaining the philosophy versus 90 hours of physical training and police control tactics added as part of the extended curriculum. It is important to emphasize that although physical training, use of force, and police control tactics are important skills, they are traditional skills that were somewhat neglected in the past and truly deserve more resource allocation in contemporary training. Nevertheless, they have to be balanced against new skills such as leadership- or integrity-related skills. It is worth mentioning that the CPD added additional hours related to the development of new skills, but when one looks at the 220 hours added by the department for the new courses, less than half (about 91 hours) are dedicated to the new skills and the remaining 129 hours to the traditional skills.

4. *What and how much is ignored?* The same set of questions that need to be addressed in point 3 is relevant here. For example, again using the CPD's newly devised training curriculum, 7 hours were devoted to military drill and formation and 1 hour to saluting. Other topics included in the curriculum can be perceived of as "disposable" as well. The answer to defenders of military drill and formation is that, yes, it can be thought of as being

extremely important, that it installs discipline and projects a professional image. However, in the era of the new skills it will be extremely hard to find another service-oriented agency that practices military drill and formation other than various military organizations instead of exposing its personnel to 7 hours of ethics or integrity. Police morality as a separate module is offered for 4 hours, although some of these themes are probably integrated under diversity management or even in the CAPS module. However, the fact remains that as a separate module, morality receives 4 hours and military drill 7 hours! When you hide the topic under a different title—if, indeed, this is the case—the message projected to the new recruit is not necessarily one of importance and relevance. When a message as important as police morality receives separate treatment under a separately titled module it is wise to make sure that it will not be contrasted against the hours devoted to drill and formation. Can we completely ignore the military drill and formation? Perhaps, but only if due to the scarcity of resources, the only choice, is between drill and a module on ethics. For the CPD, the addition of a number of hours to the topic of integrity is possible without sacrificing drill, and if such an addition is not possible, the 7 hours can be divided into two modules: 3 hours for drill and 4 hours for ethics.

Another example relates to the number of hours allocated to physical fitness. Once again, we do not wish to detract from the importance of physical fitness. Unfortunately, in the real world very few departments have mandatory in-service physical fitness programs that would continue the well-intended tradition established in the training academy. Unfortunately, given the fact that other very important topics are not covered during academy training but present real problems for police officers on a daily basis, such as lack of leadership skills, the author advocates minimizing the number of hours devoted to physical fitness and increasing the hours devoted to the new skills.

After addressing the key issues discussed above, a blueprint for an ideal academy can finally be revealed. Following are the major contours of an academy:

- Topics covered reflect a combination of traditional and new skills.
- The allocation of resources provides a true balance between traditional and new skills.
- Topics are offered in a nonsequential mode, each topic broken into smaller instructional modules and offered a number of times during the training period.
- The module topics reflect the importance of the issues addressed and are not covered under a general subheading such as "Human Behavior." A direct to-the-point heading such as the CPD's module entitled "Gays, Lesbians, and the Police" sends a clear message about where a department stands on a particular issue. The clear message is later absorbed in a nonnegotiable way by the recruit, leaving no room for interpretation.

- Textbooks are either purchased by the academy or recruits are required to purchase books relevant to the instruction topics. Books relevant to the new skills are purchased based on a thorough bibliographic search based on, and with help from, academicians who specialize in the field of police science.

- Instructors' manuals and simulation software purchased for instruction in the traditional skills are purchased based on, and with help from, professionals, members of the American Society of Law Enforcement Trainers or IDEALIST.

- The training curriculum is revised annually, and updated proactively rather than post-hoc.

- The training facility is assessed and evaluated every two or three years by an outside agency to ensure that the program offered actually provides the basic skills to the new breed of service-oriented police officer.

- Each recruit leaves the academy with clearly outlined areas in which he or she needs to receive additional training. The field training and evaluation officer or other supervisor will be responsible for addressing problematic issues during the follow-up stages.

SUMMARY

The critical issues emphasized in this chapter with regard to basic academy training can be summarized in two words: length and content. Through an overview of a number of typical academies, each representing a different model of basic academy training, from in-house, through regional, to state, the reader was presented with the existing deficiencies.

It is extremely difficult to revise any type of training, both in length and in content, without close cooperation from a state's POST commission and major support from local politicians. Nevertheless, every academy director can easily "move around" certain topics and reallocate hours. Although the mandatory minimum required by the local POST cannot be disregarded or compromised, most academies add hours of training to the minimum outlined by the POST. The commission added hours and topics should be the target of careful evaluation and consideration.

It is beyond the scope of this book to outline the minimum number of hours needed for each segment. Each agency will have to customize its curriculum based on the available resources. However, as long as the basic principles outlined in this chapter are followed, a new, advanced, and truly proactive academy training will be offered to the truly deserving recruits and the public. (See Chapter 5 for a discussion of an integrated model of academy training.)

References

Bennett, W. B., and Hess, K. M. (1996). *Management and Supervision in Law Enforcement*, 2nd ed. St. Paul, MN: West Publishing.

Bittner, E. (1980). The functions of police in a modern society. In L. K. Gaines and G. W. Cordner (eds.), *Policing Perspectives: An Anthology* (1999). Los Angeles: Roxbury Press.

Charlotte–Mecklenburg Police Department (1999, 2000). *Training Manual.*

Chicago Police Department (1999). *Training Manual.*

Christian, K. E., and Edwards, S. M. (1985). Law enforcement standards and training councils: a human resource for planning force in the future. *Journal of Police Science and Administration, 13,* 1–9.

Edwards, T. D. (1993). State police basic training programs: an assessment of course content and instructional methodology. *American Journal of Police, 12,* 23–45.

Fosdick, R. B. (1969). *American Police Systems.* Montclair, NJ: Patterson Smith.

Frost, T. M., and Seng, M. J. (1984). Police entry level curriculum: a third year perspective. *Journal of Police Science and Administration, 12,* 251–259.

Indianapolis Police Department (2000). *Training Manual.*

LEMAS Report (1996). Law enforcement management and administrative statistics. *http://www/ojp.usdoj.gov/bjs/*

Matheny, T. L. (1996). Entry level police training: evaluating the standard, a consumer's guide. Master's thesis. Jersey City State College, Jersey City, NJ.

McManus, G. P. (1970). *Police Training and Performance Study.* Washington, DC: U.S. Government Printing Office and Law Enforcement Assistance Administration.

New Orleans Police Department (1999). *Training Manual.*

Northern Virginia Police Academy (1999). *Training Manual.*

Reith, C. (1975). *The Blind Eye of History: A Study of the Origins of the Present Police Era.* Montclair, NJ: Patterson Smith.

Saint Petersburg Junior Police College (1998). *Training Manual.*

Shusta, R. M., Levine, D. R., Harris, P. R., and Wong, H. Z. (1995). *Multicultural Law Enforcement: Strategies for Peacekeeping in a Diverse Society.* Upper Saddle River, NJ: Prentice Hall.

State of South Carolina (2000). *Basic Academy Training Manual.*

Stead, P. J. (1977). Joseph Fouche: the Napoleonic model of police. In P. J. Stead (ed.), *Pioneers in Policing.* Montclair, NJ: Patterson Smith.

Suffolk County Police Academy (2000). *Training Manual.*

Thibault, E. D., Lynch, L. M., and McBride, R. B. (1998). *Proactive Police Management.* Upper Saddle River, NJ: Prentice Hall.

Walker, S. (1999). *The Police in America*, 3rd ed. New York: McGraw-Hill.

Washington, DC Metropolitan Police Force (1998). *Training Manual.*

Wroblewski, H. M., and Hess, K. M. (1997). *Introduction to Law Enforcement and Criminal Justice.* St. Paul, MN: West Publishing.

Field Training
Officer Programs

Introduction

A "doer's" contempt for educators, embodied in the oft-quoted one-liner, "Those who can't *do*, teach!" was taken one step further by filmmaker Woody Allen who wrote, "And those who can't teach, teach gym!" Law enforcement personnel are "doers." Perhaps this is why there long existed among line personnel an occupational bias against education and, by extension, against training.

For many years, the sentiment prevailed among law enforcement officers that a college education had no relevance to work. Indeed, college-educated officers often were presumed to be "soft" until they proved otherwise. Law enforcement academy training, tolerated as a necessary career prerequisite, bore little relation to one's experiences on the job. It was to be endured; officers understood that the knowledge and skills needed to get by on the streets or in a jail could only be learned by *doing*. These sentiments have been diminished but remain today.

In all probability, anti-education and anti–classroom training bias has its roots in law enforcement personnel's perceptions of the uniqueness of their work. As discussed earlier, many law enforcement officers believe that citizens cannot possibly understand the demands, stresses, and nuances of their jobs. Many are unconvinced that managing police officer encounters with members of the public are subjects that are amenable to conventional classroom training. Most doers believe that watching a veteran officer intercede to defuse a domestic dispute or to reduce tensions in a jail dayroom is the only way to learn how to do it.

Ultimately, a rookie learns by doing, and doing means following the example set by veteran officers. This technique is reminiscent of the early European guild concept of apprenticeship. Given the value that the law enforcement community attached to learning by doing, it is curious that scant attention was paid to the qualifications and skills of the veteran doers from whom the rookie was to learn the skills of his or her craft. Not until the early 1970s was a credible model developed and implemented with the degree of success that allowed other agencies to emulate it. That model was the San Jose, California, Police Department's Field Training and Evaluation Officer (FTO) Program.

THE SAN JOSE FTO MODEL

Begun in 1972, the San Jose model FTO program is the most widely recognized program of its kind. It is emulated by law enforcement agencies throughout the country that are seeking to achieve several key organizational objectives, primarily improvement of recruit officer training. The San Jose program was developed in response to an incident that occurred in the early spring of 1970. A recruit police officer was involved in a serious traffic accident while negligently operating a police vehicle on duty. A passenger in the other vehicle was killed and the officer was seriously injured. The officer was dismissed by the city. A review of his personnel record revealed serious inadequacies in the department's recruit training and evaluation procedures. The incident led San Jose Police Department managers to conclude that the process by which rookie officers acquired practical on-the-job knowledge and skills needed to be formalized. What emerged from San Jose's experience was the concept of the FTO program.

The original San Jose FTO model is a mentoring type of model, designed to provide a practical "information bridge" from the training academy to the job. San Jose's model is 14 weeks long: three 4-week periods of training, followed by a 2-week evaluation period. Upon graduation from the training academy, a recruit is assigned to the FTO program. Training occurs on each of the three workday shifts and is concentrated in two districts. The FTO team working with the recruit-trainee plans days off and study time well in advance. The recruit-trainee is assigned to a different FTO for each training period. This enables the trainee to be exposed to various styles of police work and to ensure that he or she will not be penalized because of a personality clash with a single FTO (Cox, 1996, Johnson, 1995). Each FTO provides specific training consistent with a curriculum developed for the program; each FTO evaluates the trainee's abilities and developing skills and knowledge.

FTO programs have been implemented by hundreds of law enforcement agencies across the United States. The original San Jose program has been modified by many agencies to suit their own needs and objectives; many other agencies have developed their own unique programs.

WHY AGENCIES DEVELOP FTO PROGRAMS

For police departments, an important incentive for implementation of an FTO program is accreditation. The Commission on Accreditation for Law Enforcement Agencies (CALEA) requires that all police agencies seeking accreditation have a formal FTO program. But there are other important reasons for agencies to establish FTO programs.

Reporting results from a 1986 national survey of more than 300 responding agencies, McCampbell (1986) documented the reasons most frequently cited by the 188 agencies that reported having established FTO programs:

- The need for standardized training of recruits
- Personnel needs such as standard evaluation techniques and regular evaluations by FTOs to validate agency hiring, retention, and termination decisions
- Reduction of civil liability complaints against agencies and personnel

Molden (1987) concurred with McCampbell, citing four potential benefits to agencies that establish FTO programs:

- A structured, standardized learning experience for recruits in preparation for solo patrol
- The transfer and application of classroom training to real problems and situations encountered on the job
- The importance of exposing a recruit-trainee to an FTO who serves as mentor, guide, advisor, and role model
- Documented evaluation of recruit performance to validate selection procedures, inform retention and termination decisions, defend against false equal employment opportunity and liability charges, and determine readiness of officers for solo patrol

Molden (1987) theorized that recruits who participate successfully in an FTO program will, as a consequence of having been trained under standardized conditions, have improved self-image, perform better, and be better able to contribute to the safety and welfare of citizens. Conversely, incompetent, ill-suited candidates will be discovered and terminated, thereby reducing agency liability. Discrimination and liability charges can be defended successfully with records generated during the training cycle. In response to the discrimination provisions of Title VII of the Civil Rights Act of 964, all requirements in the job selection process must be shown to be job-related, and there is nothing as job-related as the job itself. Properly structured, an FTO program will measure the candidate against job standards in a very realistic setting. Standardized training and evaluation enable agencies to better predict performance outcomes.

Some agencies that do not have FTO programs attempt to expose recruit-trainees to "real-life" on-the-job experiences through simulation. Trainees are shown films of searches; academy instructors stop the films at key points to emphasize good and bad techniques. But something important is missing from the training: the environment on the street itself. The extent to which the presence of a real-life suspect distracts a rookie officer as he or she performs a search cannot be calculated. Some officers will be so influenced by the tension inherent in any new environment, particularly one as potentially dangerous as an encounter with an armed (or perceived to be armed) suspect, that academy-taught lessons are quickly forgotten. Other rookies are unperturbed and implacable.

Despite the anecdotal contention that field trainers actually encourage recruits to forget what they learned at the academy, the formal goal of an FTO program is to help a trainee to be able to apply academy lessons in the work environment—on the street. Proxemics, the study of space as we perceive it and use it around us, is a concept that should emphasized and incorporated into FTO programs so that trainers will remain aware of the importance of the work environment to the learning process of the recruit-trainee in his or her charge.

Proper theoretical security techniques can be *taught* and *understood* in an academy; however, realistic security techniques may be *learned* and *absorbed* only during real working conditions. Expressed another way: Academy training can illustrate how things are supposed to be; an effective FTO program can demonstrate how things are.

KEY TO SUCCESS: SELECTION, TRAINING, AND RETENTION OF FTOS

Webb and Hendrix (1987) mailed surveys to 450 police agencies and received responses from 143 agencies of all sizes. The average size of the departments responding was 466 officers Of the agencies responding, 129 had field training officer programs, with an average of 31 FTOs per department. Of the 129 departments with FTO programs, 112 provided their FTOs with training. The average number of hours provided was 120. The minimum number of hours provided was 2. As noted earlier, a department may "customize" its FTO program to derive maximum job-specific skills acquisition and retention for its officer force.

A key component of any prospective FTO's preparation to assume the weighty responsibilities with which he or she is entrusted is substantive preparation in the rules and regulations by which daily agency operations are carried out. Agency policies and procedures must be reviewed; revisions must be studied and understood, as must changes in legal mandates. Although it is assumed that FTOs are knowledgeable in agency procedures, many agency program coordinators build into FTO training sufficient time for comprehensive review.

For many officers, working as an FTO is the first time the officer assumes formal responsibility for evaluating someone else's performance. Because evaluations are the measures of a recruit-trainee's progress and the determinants for whether he or she will be accepted into the agency's officer ranks to begin a career in law enforcement, the evaluations are extremely important.

Some prospective FTOs are taught during the training period to be unbiased, impartial, and fair in their evaluations. Personal dislike for a recruit-trainee cannot be allowed to influence an evaluation. However, not all FTOs undergo specially designed training; many just receive written material to study on their own. In the latter case, it is hard to envision exactly how any FTO overcomes personal biases and prejudice. Similarly, "human" pressures to allow a recruit-trainee who is not meeting minimum standards to "pass" are difficult to resist. People do not like to see themselves as "causing" another's failure. FTOs presumably understand that a recruit-trainee's failure is not the FTO's ultimate "fault." The reality is that some recruits manage to "slip by" the recruitment and hiring process even though they are, in fact, unsuitable. By identifying unsuitable officer candidates, FTOs promote safety and security for all officers and for the public and reduce the risk of agency liability for wrongful hiring and retention.

FTO PROGRAMS OF THREE POLICE DEPARTMENTS

Given the importance of the FTO programs in the decade of liability suits against police departments across the United States, Haberfeld (1998a–c) looked closely (as opposed to survey methods) at three police departments, varying in size and location, and their implementation of this crucial part of police training 26 years after the inception of the first program. Implementing an FTO program is not without problems. The way in which three departments that vary in size and resources do so illustrates and highlights the problematic issues involved in implementation of law enforcement training. Since this book is intended to serve not only as a traditional textbook that describes events and phenomena factually, but also as a tool for the implementation of effective training for agencies that differ in size, location, and resources, the agencies presented below will be identified simply as agencies A, B, and C. This type of labeling is used to eliminate any possible obstacle in implementation of a given model based on bias toward a given location, as we believe that the models presented can be imported and modified by any law enforcement agency.

Agency A: 1500 Sworn Officers

After graduation from the police academy, the probationary officers are assigned to a field training officer in one of the patrol districts. The length of training is 12 weeks, divided into three phases. The probationary officer is assigned to three FTOs, each of whom will train the recruit for the duration of one phase.

Phase I. Phase I lasts for 4 weeks, during which the probationary officer's performance is evaluated on a daily basis. During the first week the recruit observes the FTO without interference. Starting in the second week, the FTO has full discretion to decide when and in what capacity the recruit could interfere. As soon as the FTO decides that the officer has the ability to perform, he or she is allowed to engage in one of the 15 tasks assembled for the purpose of performance appraisal. Probationary officers are graded based on their ability to perform each task, ranging from "exceeds," the best performance, through "good," acceptable, to "needs development," the failing grade. The two officers are classified as a "one-officer unit"; they are not allowed to answer a call that requires two officers.

Phase II.   Phase II lasts 4 weeks. In essence, phase II does not differ much from phase I, the assumption being that during the first 4 weeks the probationary officer cannot possibly be evaluated on each of the 15 tasks. Therefore, whatever was not evaluated during phase I is stressed during the second period. In addition, the recruit is assigned to a new FTO and an attempt is made to place the officer on a different shift than the one to which he or she was assigned during phase I.

Phase III.   Phase III lasts 4 weeks. This phase is divided into two periods. During the first 2 weeks the probationary officer is still evaluated and graded on the ability to perform 15 tasks. The recruit is assigned to a new FTO and a different shift (if possible). During the last 2 weeks of phase III, officers are expected to be able to handle every task and any other situation immediately and correctly. During all phases, FTOs evaluate and grade recruits using the same form and the same performance appraisal format. To complete the program the probationary officer must demonstrate the ability to perform "good"-level work in all rating areas. Probationary officers who are unable to attain this level are recommended for remedial training at the end of phase III.

Final Evaluation Format.   The final end-of-phase evaluation format is presented to the review board in the training division. The review board is comprised of the training captain, who is the director of the police training academy, FTO sergeants, and a major in charge of the patrol district to which the probationary officer was assigned. Upon completion of the review, the recruit is assigned to one of the patrol districts and remains on probation for the remainder of the probationary period (a year and a half from the day of hire). At the end of the FTO program the recruit evaluates all the FTOs assigned to him or her during the three phases.

Agency B: 500 Sworn Officers

Agency B's FTO program is based directly on the one developed by the San Jose Police Department. Fifteen years ago, police officers from San Jose came to Agency B to teach local officers how to operate their program. The original program was modified a number of times; today, it consists of four phases.

Phase I.   Phase I lasts 28 days, during which the concentration is on training (90 percent of the time), and a little bit of evaluation (about 10 percent). The criteria for evaluation are either + or − . The first phase is characterized by high liability; the officers are supposed to "get their feet wet." The high liability is a product of potential complaints that could be filled by citizens dissatisfied with the service provided by the probationary officer after completion of the FTO program. The FTOs try to document everything that is been taught. The liability may come not only from outside but also from officers themselves—claiming that they were not trained. One FTO assigned to this period will also be the FTO to whom the probationary officer will return for the last phase (phase IV) of the training cycle.

Phase II.   Phase II lasts 28 days. The probationary officer is assigned to a different FTO and a different shift in a different area of the city. The city is divided into three districts and the probationary officer is introduced to all three during the FTO training. This phase is characterized by 50 percent training and 50 percent evaluation. The officers are assigned to more "real work"; there is also more responsibility. The evaluation form differs from the one in phase I; this time the categories for evaluation range from most to least acceptable using a numerical rating. The evaluation itself is focused on the behavior and performance of the officer. This is a "solo performance," an officer operating in the city, trying to balance things out, with the FTO looking for positive and negative characteristics. If the probationary officer shows deficiencies, phases I and II can be extended for a varying amount of time.

Phase III.   Phase III lasts 28 days. The majority of officers usually make it to phase III, which is characterized by 90 percent individual work, or 90 percent evaluation and 10 percent training. Again, the probationary officer is assigned to a new FTO. Frequently, there is an orientation day during which the probationary officer learns the new streets and new area to which he or she was assigned. During this stage, FTOs are usually able to catch those who need an extension period or who fail the FTO program. FTOs meet on alternate Fridays with the FTO sergeant, and sometimes the FTO lieutenant, to discuss the results of the training and issues concerning the probationary officers.

Phase IV.   Phase IV lasts 2 weeks. During this phase the officer is reassigned to the first FTO (the one who trained the officer during phase I). The FTO rides in the car in plain clothes. This phase is considered to be for evaluation purposes only. The car is considered to be a "one-officer car," and the purpose is to find out whether or not the officer is doing a good job. The probationary officers are expected to take total charge of any situation. After completion of phase IV, the officer is introduced to an informal stage of the FTO program (developed by Agency B), also known as phases V and VI.

Phase V.   Phase V lasts 2 weeks. The officer rides with the traffic division and investigates accidents, especially DWI (driving while intoxicated) investigations.

Phase VI.   Phase VI lasts 2 weeks. The officer is assigned to a community-oriented officer or CPO in the area in which he or she is going to be assigned permanently. During phases V and VI, the probationary officers are not evaluated by the FTO; they are assigned to regular traffic officers, or CPOs. The evaluation form is a generic form, submitted by the traffic officer or the CPO to the FTO unit once a week. Both phases can be postponed, depending on the availability of resources and/or personnel (Table 5–1).

Final Evaluation Format.   FTOs have to complete a daily observation report. The FTO sergeant reviews the form with the probationary officer, his or her sergeant, and the administrative sergeant before sending it to the FTO lieutenant. The FTO sergeant has different types of work responsibilities within the unit. The administrative sergeant, one of the regular FTO sergeants, coordinates the paperwork assigned to the FTOs. On alternate Fridays the probationary officers receive additional training, called *hands-on training*: for example, a lecture from an officer assigned to the K-9 unit. The recruits are evaluated during these Fridays as well. For a probationary officer to be able to perform solo with the full capacity of a sworn officer (in police jargon, to be "cut loose"), the FTO must submit a report to the FTO sergeant, who, in turn, makes a recommendation to the FTO lieutenant. The final decision is made jointly by the FTO lieutenant and FTO sergeant. Finally, the FTOs themselves are evaluated by their probationary officers. The FTO does not see this evaluation until the officer is off probation, but the FTO sergeant monitors evaluations throughout the FTO training period.

Table 5–1   FTO Phases and Stages

Phase I	28 days 90% training/10% evaluation FTO 1
Phase II	28 days "Solo performance" FTO 2
Phase III	28 days 90% evaluation/10% training FTO 3
Phase IV	14 days "One-man car" Back to FTO 1
Phase V	14 days Traffic division No FTOs
Phase VI	14 days "C-O-P" No FTOs

Agency C: About 320 Sworn Officers.   Agency C has instituted its field training and evaluation program with two purposes in mind: to identify specific weaknesses or deficiencies of the probationary officers under actual field conditions and to develop remedial training programs to address and correct the deficiencies, and by continual evaluation, to identify and remove those persons who are unable to attain the level of proficiency expected of an officer serving in Agency C. After graduation from the state training academy, the probationary officers are assigned to a field training officer in the uniform patrol division. The length of training is 12 weeks, divided into four phases. At the end of weeks 2, 6, and 10, the probationary officer will rotate to a new FTO, ending with the original FTO for final evaluation.

Phase I.   This training and evaluation phase lasts 2 weeks. The first week, called "Orientation Week," familiarizes the recruit with the new surroundings. The recruit observes the FTO and may be allowed to assist in some cases, based on individual abilities. Training and evaluation begin on the first day of the second week and continue to the end of phase III. The status of the two is considered to be a "one-officer unit," and a new recruit is not expected to perform competently as a backup at this time. The FTO completes a daily evaluation report and develops and institutes remedial training where needed. When rotation occurs, the FTO completes an end-of-phase evaluation report, which is forwarded through the chain of command for approval and is then placed in the probationary officer's training file.

Phase II.   Phase II lasts 2 months and is also considered to be a training and evaluation phase. During this phase it is up to the recruit to demonstrate his or her abilities without the FTO's leadership. The FTO completes a daily evaluation report. Again, if needed, remedial training is instituted to correct the problem. At rotation, an end-of-phase report is completed and processed, as in phase I. If at the end of phase II it becomes necessary to extend the phase, with the training division's approval a probationary officer is granted an extension and is assigned on the basis of what shift and which FTO would best accomplish the goal of extension. During an extension period, daily and end-of-phase evaluation reports are completed.

Phase III.   Phase III lasts 2 weeks and is considered an evaluation period. The recruit should be able to act on his or her own without help from the FTO. The FTO maintains supervisory control over the probationary officer and is responsible for his or her actions. During the last week of evaluation, the FTO dresses in civilian clothing to minimize public contact. Daily evaluations are completed during this period. To complete the program, the probationary officer must demonstrate the ability to perform "4-level" work in all rating areas (the rating scale is numbered 1 to 7, with 1 being "not acceptable," 4 being "average," and 7 being "superior"). Probationary officers who are unable to attain this level are recommended for termination of employment.

Phase IV (Final Evaluation Format).   After completion of the field training program, the recruit is assigned to regular patrol duties. The probationary officer is assigned to a "one-officer-unit" patrol, where he or she is closely evaluated by shift supervisors. At the end of each month, a supervisor's monthly evaluation report is completed by the shift supervisor and forwarded through the chain of command to the training division. These monthly evaluation reports continue until the probationary period (24 months) ends.

If the recruit is transferred to the traffic division for a 2-week period of training, at the completion of each week the accident investigator completes a traffic section weekly report and forwards it to the training division. The final decision to retain a recruit is made by the shift supervisor, team lieutenant, and the training division after reviewing the end-of-phase report. Upon completing each phase of training, the probationary officer writes a brief evaluation of the training received. The field training critique form is distributed to recruits at the end of training. The probationary officer returns the completed form to the training division. The form becomes a part of the FTO's file only after the probationary officer completes the training phase successfully.

PREFERRED MODEL, SELECTION CRITERIA, AND INCENTIVES

FTO programs help compensate for selection errors and inadequate academy training. They provide role model for recruits. They allow for application of knowledge and skills that cannot occur in a classroom. Prior to discussion of the ideal models, the potential hazards of existing FTO programs should be addressed. The International Association of Chiefs of Police urges that certain potential hazards be recognized and avoided (Bennet and Hess, 1996):

1. Overemphasis on technical skills
2. More evaluation than training
3. Typing of recruits
4. Too short and/or too demanding programs
5. Too young and/or too inexperienced FTOs
6. Disliked versus incompetent recruits

The vital importance of FTO programs is documented by Molden (1993) as he describes the investigation of the Los Angeles Police Department's alleged use of excessive force. He suggests several generally accepted questions to ask to determine if an FTO program will be effective:

1. Is the program supported by management?
2. Are FTOs selected properly?
3. Are FTOs compensated?
4. Are FTOs trained properly?

5. Are accurate and detailed records maintained?

6. Is the program organized within the chain of command with designated supervision and reporting responsibility?

7. Is a field training guide used? Is it based on a job/task analysis?

8. Do recruits receive sufficient field training time before being assigned to solo patrol?

9. Do FTOs maintain contact with the training academy curriculum and staff?

10. Is there rotation among field training assignments?

11. Are there guidelines for evaluation of recruits?

Molden (1993) addresses one very importune aspect of the FTO program, the field training guide. According to his specifications, the guide should be organized around the following principles:

1. The guide should contain a complete list of the skills, knowledge, and abilities to be taught to a recruit and those needed for safe and efficient performance of solo duty.

2. The guide should be organized into general categories such as patrol, investigations, and vehicle stops.

3. The general categories should be broken down into subcategories and then further into specific learning tasks.

4. Each topic in the guide might be followed by three columns: explain, demonstrate, and perform.

Two separate topics of importance emerge after the first set of problems and recommendations is addressed. In addition to the format and to problems with the programs themselves, the selection criteria for choosing the FTO and the incentives/benefits offered trainers seem to be of critical importance. Based on research conducted in three agencies, Haberfeld (1998a–c) concluded:

1. The selection criteria as well as benefits/incentives differed significantly in the three departments.

2. In Agency A the prospective FTO faces a selection committee comprised of the training captain and a representative (usually, a major) from each patrol district. The candidate has to be recommended by both his or her first-line supervisor and captain. The selection criteria include:
   - History of disciplinary violations
   - Record of attendance and punctuality
   - Ability to communicate (both oral and written)
   - Rating of at least "good" on performance appraisal (for at least two years prior to the date of review)

- Demonstrated commitment to departmental mission statement
- Must have served at least three years with the department (no maximum length of service to disqualify a candidate)

After selection the candidate must complete a 40-hour training session at the academy.

3. In Agency B the selection process begins when a general notice is send around of an opening. The selection criteria include:
   - History of disciplinary violations (no disciplinary charges filed against him or her in the past 18 months)
   - Must be at least three years of probation
   - The skills and training desired are listed in general order and vary depending on need.

After selection the candidate must go through a 1-week training course, a combination of academy and in-house training. If necessary, the prospective FTO can be sent back for additional training in certain areas.

4. In Agency C an FTO is selected by a team commander and his or her supervisor. The selection criteria include:
   - The candidate must have at least two years of service. (In reality, however, due to high turnover, officers who are a couple of weeks or even months short of the two-year requirement become FTOs while either still on probation or barely out of the probationary period.)
   - There is no maximum length of service that would disqualify a potential candidate.
   - No formal schooling or training is required.

An official description in the form of a training manual which provides guidelines for prospective candidates is issued for the purpose of self-study.

The incentives or benefits to becoming an FTO are similar in all three departments, even though the monetary compensation differs significantly. In general, it can be stated that in all three departments the following sentiments expressed by the FTOs hold true with regard to nonmonetary incentives:

1. FTO status looks good "on paper," for the purpose of promotions.
2. Even though there is no formal status assignment, some people volunteer because of the admiration they had for their former FTOs.
3. Initially, the officers felt committed to the program—"once an FTO, always an FTO." Today, an FTO has the option of terminating a contract and is therefore given more discretion over the continuance of his or her participation.
4. Internal politics play a role. If somebody is asked to become an FTO, especially during a shortage of candidate's time, a positive response might generate a return favor in the future.

5. Sharing information the younger and new officers, shaping and molding them, gives back some of the training that they themselves received from FTOs.

On the other end of the spectrum, Molden (1994) identified the following as common complaints of FTOs:

1. Money
   - Lack of compensation
2. A one-officer car
   - When short-handed, the department uses an FTO car as a two-officer unit.
   - The department should have a clear policy that says: The FTO car is a one-officer car—provide appropriate backup for potentially hazardous situations.
3. Ignoring FTO recommendations
   - The frequently heard excuse: We invested too much money in the recruit and we cannot afford to fire him or her.
   - We might get sued for unlawful termination.
4. Poor writing
   - Probationary officers exhibit poor writing skills.
5. The daily observation report
   - Excessive paperwork is involved in daily evaluations.

Earlier, Molden (1990) addressed the topic of FTO burnout. The FTOs carry a double load: as a patrol officer and trainer. He or she also constantly has the weighty responsibility of another officer's behavior and safety. It is logical to believe that on average, burnout is a more critical problem for FTPs than for officers. The following suggestions are put forth by Molden as a self-help guide for FTOs:

1. Keep the work interesting.
   - Assign the FTO an interesting beat, assignment, or shift.
2. Give recognition.
   - Provide monetary rewards in addition to base salary or overtime pay.
   - Give priority status to assignment or desirable training classes.
   - Provide a take-home vehicle.
   - Issue a distinctive badge or patch.
   - Arrange media publicity.
   - Assure career path advancement.
3. Provide R&R.
   - Provide sufficient break time between assignments.
   - Except on a short-term emergency, an FTO should probably not train more than two or three recruits a year.

4. Avoid "other" duties.

   ▪ No one can be expert in everything. A field training assignment by itself is a demanding job requiring the full attention of a dedicated, well-trained person.

   ▪ Do not assign the FTO to other responsibilities.

5. Limit the assignment.

   ▪ Limit the FTO assignment to two or three years.

   ▪ If the FTO was selected properly, he or she is probably ready to move up to another challenge.

A successful program that addresses the burnout phenomenon, coupled with an attractive benefit package, still cannot ignore the crucial importance of the initial selection of adequate FTOs. An attempt made by the Indianapolis Police Department to deal with this issue is worth attention. The Assessment Center process is designed to identify and select patrol officers best suited to perform as field training officers. This process utilizes exercises that are structured to simulate actual job requirements of a field training officer. During the assessment center process, trained assessors observe and evaluate the applicant's performance in order to determine which applicants demonstrate the greatest potential of becoming successful training officers.

The Indianapolis Police Department has developed six exercises for use in the field training officer assessment center process (Decker, 1990):

1. Situational exam (in-basket)

   ▪ The applicant is required to assume the role of a newly assigned field training officer.

   ▪ The applicant is presented with a series of situations.

   ▪ After taking appropriate action, the applicant is interviewed about the approach chosen.

2. Individual presentation

   ▪ The applicant is required to make a presentation regarding a specific issue (controversial topics related to field training).

3. Written letter

   ▪ The applicant is required to prepare a letter to the assessment board. Written communication skills are measured.

4. Oral interview

   ▪ The interview is designed to identify how the FTO applicant views his or her responsibility toward the FTO program.

5. Role-playing/counseling session

   ▪ Communication skills, judgment, problem analysis, sensitivity, and planning are measured.

6. Competency test
   - A written examination designed to measure the applicant's knowledge of department policy, rules, and regulations is administered.

There is, of course, the desired picture of what the FTO process should look like as opposed to the grim reality of resources, specifically personnel and time. A number of solutions to these problems are discussed in Chapter 9. For now, a recommended model for FTO training is presented which represents an attempt to reflect the possible rather than the ultimately "desired."

The attached models represent two schools of thought. The first model, "FTO Phases/Stages" (Table 5–2), is composed of six FTO phases that should ideally be included in each FTO program. The second model, "Probationary Period" (Table 5–3), advocates for inclusion of the FTO period within the training academy. The primary reason is the need to integrate the theory with practical knowledge in the field. This model is especially relevant for in-house academies, since they can easily customize their FTO program to the work environment outside the academy. It is, however, even more important, although more difficult, in the case of state and regional academies. They would have to coordinate this model with every agency that sends its recruits for training in the regional or state facility, but this can be accomplished given the importance of "real-life"

Table 5–2   FTO Phases and Stages

Phase I	28 days 90% training/10% evaluation FTO 1
Phase II	28 days "Solo performance" FTO 2
Phase III	28 days 90% evaluation/10% training FTO 3
Phase IV	14 days "One-man car" Back to FTO 1
Phase V	14 days Traffic division FTO 1
Phase VI	14 days "C-O-P" P.O.P. FTO 1

Table 5–3   Probationary Training

Phase I	2 weeks In-force familiarization Introduction to policing
Phase II	20 weeks Academy training Student-centered, interactive, participative, skills-based application of law
Phase III	12 weeks FTO—Part I Phases 1, 2, and 3
Phase IV	2 weeks FTO—Part II Phase 4
Phase V	2 weeks FTO—Part III Phase 5
Phase VI	2 weeks FTO—Part IV Phase 6

testing prior to releasing the recruit to a given department that does not have the adequate resources, especially the personnel, to put the recruit through the necessary FTO program (e.g., Agency C in this chapter). Many examples can be found in small police departments, with 50 officers or less. These small departments represent the majority of police departments in the United States. Over 95 percent of police departments in the United States will at some point experience a shortage of personnel to run an adequate FTO program. For this reason, mandatory inclusion of FTOs within academy training will afford every recruit with the necessary skills on both theoretical and practical levels. It will also allow for standardized training concepts, which will definitely further the professionalism of police work.

The title of the Probationary Period model refers to six FTO phases that are described in the first model under the heading "PTO Phases/Stages." The six FTO phases are based on the model utilized by one of the agencies described in this chapter (Agency B) but changes the roles of the FTOs assigned to each and every phase. The Probationary Period model is based on ideas introduced by British police, whose model of training incorporates the academy with the preservice and FTO program. Implementation of probationary training in state and regional academies is discussed further in Chapter 15, where the size of the department will be matched with the available resources.

SUMMARY

The goal of this chapter was to provide some information with regard to a very critical aspect of police training, the field training officer program. After looking at the deficiencies outlined in the academy training, the desire to improve those shortcomings becomes possible with the introduction of the FTO program. Despite the more student-centered and individualized approach to training, a number of problems were identified through a closer look at three police agencies and their FTO programs.

Similar to the picture portrayed in Chapter 4, the major deficiencies can be summarized in two words, timing and evaluation. The proposed FTO models integrated into the basic academy training emphasize the criticality of proper evaluation, based primarily on the timing of exposure to real-life policing. Merging the theory with practice has a long history of success in countries such as Japan and Ireland. A number of basic academies in the United States include their FTO programs within their basic modules, but the majority have chosen to ignore this important principle of synthesis and integration of theory with practice following a very thorough evaluation.

REFERENCES

Bennet, W. W., and Hess, K. M. (1996). *Criminal Investigation*. St. Paul, MN: West Publishing.

City of Charleston, South Carolina, Police Department (1996). *Field Training and Evaluation Manual*.

Cox, S. M. (1996). *Police: Practices, Perspectives, Problems*. Needham Heights, MA: Allyn & Bacon.

Decker, K. H. (1990). Security management, *Arlington, 34*(1), 44.

Haberfeld, M. (1998a). Confronting the challenges of professionalism in law enforcement through implementation of the field training officer program in three American police departments. *Turkish Journal of Police Studies, 1*(2).

Haberfeld, M. (1998b). Field notes, City of St. Petersburg, Florida.

Haberfeld, M. (1998c). Field notes, City of Charleston, South Carolina.

Haberfeld, M. (1998d). Field notes, City of Charlotte–Mecklenburg, North Carolina.

Johnson, G. (1995). San Jose: a commitment to excellence. *The Trainer, 3*, 31–38.

McCampbell, M. S. (1986). Field training for police officers: state of the art. Washington, DC: U.S. National Institute of Justice, "Research in Brief" series.

Molden, J. B. (1987). Training field training officers. *Law & Order, Wilmette, 35*(1), 22.

Molden, J. B. (1990). FTO training alternatives. *Law & Order, Wilmette, 38*(7), 16.

Molden, J. B. (1993). Training to overcome accident liability. *Law & Order, Wilmette, 41*(5), 13.

Molden, J. B. (1994). Five common problems: FTOs everywhere have similar complaints. *Law & Order, Wilmette, 42*(9), 21.

Webb, F., and Hendrix, B. G. (1987). Field training pay: what a national survey found. *Training Aids Digest, Springfield, 12*(9).

INSTRUCTORS AND EDUCATORS
Where to Start?

INTRODUCTION

In this chapter we deal with qualifications, standards, and what is required of the people responsible for the training and education of law enforcement officers. The ideal and, in essence, theoretical qualifications for instructors and educators are discussed and contrasted with the real-life requirements imposed on them in the field. The reader will have the opportunity to compare the desired with existing requirements and to decide how many of the desired skills are actually found in the men and women put in charge of the critical task of preparing people who have the power to use force and perform other activities that can affect another person for the rest of his or her life.

The main purpose of the chapter is to highlight the irresponsible approach to the selection of trainers and educators in the field of law enforcement. The tasks placed on the shoulders of those teachers and the responsibility they take upon themselves are tremendous. Nevertheless, their preparedness for these tasks is minimal. Many of the skills required are not relevant to the topics being taught, and the people in charge of their selection do not have a clear idea of the qualifications of an ideal instructor. The reader is also encouraged to contrast a number of teaching methodologies (discussed in Chapter 3) against the skills and responsibilities of the academy or the FTO trainers.

One can only imagine how professional and well educated our medical doctors would be if their training and education were placed in the hands of people who have no idea how to teach, what to teach, or have little understanding of the topics they are supposed to convey to students. The medical profession as a whole would be outraged and the public would be devastated. After all, medical doctors deal with people's lives—it is a matter of life and death. Well, it is a matter of life and death with police officers as well.

It is extremely important to emphasize from the onset that there are many, many skillful, and dedicated instructors and educators in the field of law enforcement and the author does not imply that all the people involved in police training are not qualified to do so. However, even the best and most motivated instructors might be teaching what they know best without having a "bigger picture" of what their teaching might lead to. Sometimes they do not allocate enough time to a topic to be able to explore all the nuances involved. Ideas in this chapter could provide additional insight into what could be changed, added, or altered to provide the best possible solution to such difficulties.

When a firearms instructor is asked to devote a number of hours to familiarize recruits with a gun, he or she is rarely given ample time to explain the psychological and physiological impact that use of a gun has on both the person who shoots (or draws the weapon) and the person at whom the gun is shot (or pointed). Even if this is covered, it is generally discussed inadequately, leaving the recruits with a limited knowledge of the topic. In addition, the firearms instructor might be the best firearms instructor but might not be the best educator, and to explain the psychological and physiological impacts well, one needs to be as much a superb educator as an instructor (see Chapter 7).

Ideal instructors and educators in the field of law enforcement do exist, but their numbers are not adequate. Many more could be found and utilized if the people in charge of their selection would give the information presented in this chapter some consideration.

EDUCATION VERSUS INSTRUCTION

Education can be defined as developing or training knowledge and the result can be acquired by instruction or teaching (The American College Dictionary). Biehler and Snowman (1993) describe teaching as both an art and a science.

- *Teaching as an art.* Highet (1957) identified three elements that he considered to be important in successful teaching. The two first elements are *emotions* and *values* because these components cannot be "objectively and systematically manipulated." He believed that teaching is a valuable and rewarding activity, that it has to be done every possible day, that students should be excited about learning, and that "unteachable" students do not exist. The third component is *flexibility*, which refers to being able to do the right thing at the right time.
- *Teaching as a science.* Biehler and Snowman (1993) emphasized the "scientific" component as an important aspect of teaching as well, the main reason being existing research on the characteristics of expert teachers. Berliner (1986) and Shulman (1986) suggest that there are

two main areas which are important in becoming successful in the science of teaching. These are *subject-matter knowledge* and *knowledge of classroom organization and management*. When a teacher knows how to recall such relevant knowledge quickly and accurately, he or she becomes a successful expert teacher.

Good teachers appear to have a combination of artistic and scientific elements in their teaching style. If a teacher tends to focus only on scientific material, he or she is likely to come across as being rigid and mechanical. On the other hand, if a teacher focuses strictly on the need to excite and inspire students, his or her teaching may very easily be ineffective.

Instruction refers to the act or practice of instructing knowledge. It is often synonymous with tutoring, coaching, drilling and training. In contrast to an educator (who can be a college professor), an instructor is regarded as inferior in grade to the lowest grade of professor (The American College Dictionary).

Klotter (1978) identified certain qualifications as important in order to become an effective and a good police instructor.

1. The instructor needs to have extensive knowledge of the subject. This includes both field experience and knowledge of existing literature. The instructor should know more than he or she has time to teach and should be able to answer any question.

2. The instructor should know various methods of instruction and should be able to present the subject matter in a way that is understandable to the student.

3. The instructor needs leadership ability. To run a class smoothly, it must be well managed. An instructor who lacks leadership qualities will soon lose the attention of the class.

4. An instructor needs to show a professional attitude. Attitude influences student attitudes and morale. The attitude should be friendly but professional.

5. The instructor needs to show sincerity. If this is not provided by the instructor, the student will lose both confidence and interest.

6. An instructor should show enthusiasm. This is catching, and it prevents planned lessons from failing in their objectives.

7. The instructor should demonstrate salesmanship. It is important to sell the necessary material to the student, and for this to happen the student must believe in the material being taught.

In addition, a successful instructor should have a sense of humor, a pleasant appearance, and the desire to do the job.

To convey knowledge to students successfully, an instructor has to make a sensible plan. This is a crucial part of the instructional process and may be organized in the following way:

- *Preparation.* This is the planning stage of instruction—how the instructor will approach students. The material should be adapted to students' backgrounds and needs. The instructor must remember to present the material so that every member in the group learns the essential procedures.
- *Presentation.* Now the actual training begins. This element should preferably consist of three parts (however, all three are not essential for every session).

    (1) *Introduction.* Here the instructor gets the attention of students and makes sure that they are ready to receive the material that is about to be presented to them.

    (2) *Body.* This is where the material is presented. In this phase the instructor should also show how particular procedures are performed. Different methods an instructor can use are lecture, conference, demonstration, or a combination of these.

    (3) *Closing.* This is where the instructor reviews the material and clears up any questions. The student should feel that he or she has accomplished something.

- *Application.* This is where the student has an opportunity to apply the procedures and principles that he or she has learned.
- *Evaluation.* Here the instructor should check and evaluate the newly learned knowledge of the student and how the student applies it.
- *Critique.* A critique should follow the evaluation no matter how a student has performed. Students must be shown where they have failed and where they have performed satisfactorily.

These phases do not apply in every circumstance; sometimes it may be necessary to omit one or several steps. In addition, if the steps are followed, it is important that the instruction not become artificial or stereotyped. However, if the instructor considers these variables in the planning, he or she is likely to be a success.

MATURITY THEORY

Maturity theory attempts to provide an explanation of how certain people are made for some jobs and how they naturally develop to match the correct traits needed for a specific job. It is the question of "born or made." Since in this chapter we are tackling the idea of ideal police instructors and educators in the academy, FTO and advanced or specialized training settings, the concept of being born or made to perform these tasks seems especially relevant. The idea of being mature enough to perform a certain job or task has been studied on both the organizational and individual levels. In addition, the concept of maturity is addressed further in Chapter 8.

Methods of instruction, training modules, and updated topics do not constitute all the ingredients necessary for the delivery of effective training to police officers or to any other group of learners, for that matter. After taking a closer look at the qualifications and standards required of police instructors and educators, a review of the maturity theory is presented. The basis for choosing this particular theory is the author's conviction that teachers and instructors are actually born and not made, and some support for this hypothesis may be found in an analysis of the maturity theory.

Argyris (1957) describes two components that are involved in determining the issue of maturity: formal organization and individual organization. By *formal organization* he means how certain external processes are involved in the developing stages of a person. *Individual organization* refers to a person's internal processes. The two components interact to reach the best possible outcome. For example, if a basketball player follows the formal strategies for a play, he or she should, theoretically, score on each similar play. However, the play meets opposition in the person's internal feelings, which often are in conflict with the outcome of the play.

A minimum of informal organization is necessary if formal organization is to retain optimum expression of its demands (e.g., the play in a basketball game) and if the person does not want to build up too much tension. If tension builds too high, the person will lose his or her efficiency. According to Bakke (1950), formal and informal activities constitute total organization. The systems are not separable, and they are continually being modified by the people involved.

Originally, the concept of having the capacity to set high but attainable goals, willingness and ability to take responsibility, and education and/or experience of an individual or a group was referred to as *maturity theory*. The theory was based on the premise that some people are readier than others to perform certain jobs, and further, that some people may never develop the skills necessary for particular jobs. For example, a particular police officer may be excellent in directing traffic, but his paperwork has to be supervised by a superior officer (Hersey and Blanchard, 1977).

Following this concept, some police officers may be excellent police officers, based on their ability to perform various police tasks, be physically fit, and even possess the necessary knowledge in a given topic (e.g., firearms or the social sciences) but not have the necessary skills, and furthermore, will never develop the skills necessary to serve as academy instructors or FTO supervisors.

Hersey and Blanchard (1993) developed the concept of maturity into a new theory, referred to as *readiness*. The definition is similar and the main focus of the concept is still on how ready a person is or can be to perform a particular task. Readiness refers only to specific situations, not to a total sense of readiness—it is important to remember that people may respond differently in different situations. Hersey and Blanchard describe two major components of readiness. *Ability* focuses on knowledge, experience, and skill/performance, while *willingness* is based on confidence, commitment, and motivation. These two components function in an interacting system in which a change in one will have an effect on the other. As a result, readiness can be divided into four levels (Hersey and Blanchard, 1993):

- *Readiness level 1:* unable and unwilling (the follower is unable and lacks commitment and motivation) or unable and insecure (the follower is unable and lacks confidence).
- *Readiness level 2:* unable but willing (the follower lacks ability but is motivated and making an effort) or unable but confident (the follower lacks ability but is confident as long as the leader is there to provide guidance).
- *Readiness level 3:* able but unwilling (the follower has the ability to perform the task but is not willing to use the ability) or able but insecure (the follower has the ability to perform the task but is insecure or apprehensive about doing it alone).
- *Readiness level 4:* able and willing (the follower has the ability to perform and is committed) or able and confident (the follower has the ability to perform and is confident about doing it).

It may appear strange that a person can go from insecure to confident and then back to insecure. However, it is important to remember that at the lower levels a follower is leader directed, whereas at the higher levels a follower becomes responsible for task direction (Hersey and Blanchard, 1993, pp. 191–192).

As a necessary supplement to readiness theory, situational leadership theory has to be added. This is presented and discussed in Chapter 8, as it is more relevant to leadership training. However, the reader is advised to apply the second part of this theory to the information presented in this chapter.

SELECTION AND RECRUITMENT OF FTO INSTRUCTORS

Regional Academy: Virginia

In Chapter 5 we discussed in general terms some of the issues related to the selection, recruitment, and retention of FTO instructors. This part of the chapter focuses on a number of agencies and their detailed processes of recruitment and selection of the academy, the FTO, and/or advanced and specialized training instructors.

The goal of this part is to analyze, point by point, the processes put in place by a number of police agencies, to assure sound selection of instructors. The pros and cons offered in no way attempt to diminish the obvious efforts of these departments to establish high standards and quality control. The objective is to provide a reality check and to substitute for the ideal but not feasible approach a realistic and pragmatic pattern that could serve as a guideline for any training practitioner. Despite the fact that the agencies chosen obviously represent departments large enough to afford good programs, an attempt has been made to factor in the differences in size of various police departments. For example, the number of years spent on the force prior to being eligible to become an FTO instructor might be a feasible requirement in a larger agency but would be a problem for a small department that does not have the same selection options. The issue of incentives can also become problematic with regard to scarce resources in one department versus abundance in another. To control for such differences, a number of generic qualities and solutions are offered.

The following basic criteria are among the qualifications to become a basic instructor at a regional academy:

- Minimum of three years of experience, with the preference given to those with greater experience
- Education beyond high school preferred
- Demonstrated abilities in both verbal and written skills
- In above-average physical condition (physical fitness screening test is conducted in the academy)
- Ability to think logically under stressful conditions
- Possession of the personal grooming habits and demeanor that will provide a proper role model to basic and in-service trainees
- No record of disciplinary action for one year prior to application
- Preferably a nonsmoker

A closer look at these basic criteria will identify a number of deficiencies that can be addressed and corrected with minimal effort. To begin with, and taking into consideration the premise that instructors and educators are born and not made since some of us will never display the desired readiness for this task, the topic for instruction has to be identified prior to any identifiers or standards for potential instructors. For example, if an academy is looking for an instructor to deliver a module in police integrity or ethical behavior, it hardly seems adequate that the instructor has no record of disciplinary action for only one year prior to the application.

Critics of this approach may argue that some high-profile criminals, especially in the field of white-collar crime, have been invited by a number of universities and colleges to deliver lectures about their experiences and the consequences of the illegal behavior. However, whereas this approach might be appropriate in a college environment, it might not be in a basic training academy. Assigning an officer with a disciplinary record to teach ethics poses an ethical dilemma in itself.

In a similar venue, the requirement to be in above-average physical condition does not seem relevant for an instructor who wants, and is otherwise fully qualified, to teach a module on diversity and sensitivity. This particular module especially, which should emphasize tolerance toward certain groups in the general population, might benefit from an instructor who is not a marathon runner. One of the problems with which our society struggles is discrimination against various disabilities, such as obesity. The misconception that obesity is something that everyone can control, and that everybody can be in above-average physical shape, is not going to be easily dispelled by the above-average physical condition of the instructor. If anything, this requirement sends the wrong message. Some police instructors with whom the author has spoken strongly endorse above-average physical condition. The arguments they present focus primarily on the proper role model that a physically fit officer conveys to recruits. Here, again, however, the "use of force" orientation discussed in Chapter 4 seems to play an important role. It is important to emphasize from the very early stages of indoctrination to police work that the professional image or professional attitude of a police officer is not necessarily exhibited by the strength of one's muscles. A professional demeanor is exhibited by an attitude of understanding, empathy, courtesy, manners, and similar factors. Physical fitness is an extremely important component of our culture—it has been emphasized for decades, for both medical and aesthetic reasons. Nevertheless, teachers from grade school through college are not required to do push-ups as part of their job application. They are hired to educate our minds and souls, not our muscles. If police officers of the twenty-first century are totally dedicated to the idea of community-oriented policing, they should be educated and trained by people whose skills to teach and deliver training are not measured by the number of sit-ups they can perform.

Large Municipal Agency: Charlotte–Mecklenburg Police Department

The FTOs are the most important link in the socialization process of recruits. As mentioned in Chapter 5, they are the ones who will "make reality out of theory." In an ideal setting, when a given police department has its own in-house academy, the recruitment and selection process can be highly selective and controlled by the training academy staff. In this case it is also possible to hand-pick the best, due to the availability of resources. Nevertheless, as the following example shows, even in optimal environments, mistakes and errors can be found and corrected if the focus of the basic criteria will change slightly.

The Charlotte–Mecklenburg Police Department is representative of a large municipal police agency and its work can be emulated by other in-house academies. The department recommends that the selection of FTOs be the joint responsibility of the training academy staff and each patrol district. The training of the FTOs is the responsibility of the training academy staff. The prospective FTO faces a selection committee, comprised of the training captain and a representative (usually a major) from each patrol district. The candidate has to be recommended by both his or her first-line supervisor and captain. The selection criteria include:

- History of disciplinary violations
- Record of attendance and punctuality
- Ability to communicate (both oral and written)
- Rating of at least "good" on performance appraisal (for at least two years prior to the date of review)
- A demonstrated commitment to departmental mission statement and policy toward community-oriented policing and problem solving

Candidates are not eligible for the process if :

- They have one sustained violation of an A category rule of conduct within three years prior to the date of the first selection exercise.
- They have two violations of A category rules of conduct within five years prior to the date of the first selection exercise.
- They have three violations of A category rules of conduct within 10 years prior to the date of the first selection exercise.
- They have three or more sustained violations of B or C category rules of conduct (in any combination) within the three years prior to the date of the first selection exercise. (Only sustained disciplinary action is considered.)
- They have been placed on performance probation at any time within the five years prior to the date of the first selection exercise.

Candidates who do not meet all eligibility standards are not allowed to participate in the process. In addition, any candidate who becomes ineligible during the selection process or during the effective life of the eligibility list is eliminated from consideration and removed from the list. Also, any active FTO who fails to meet the qualifications and standards of the program and becomes ineligible to serve as an FTO will be removed from the field training program. It is the candidate's responsibility to determine if he or she is eligible to participate in the process. In every process the final selection eligibility list is checked against the department's records, and any candidate on the list who should have declared himself or herself ineligible is eliminated at that point.

Any or all of the following testing procedures may be utilized by the FTO selection committee to assess the FTO applicants and to assist in the selection process:

- Writing and/or grammar exercise
- Oral interview
- Practical assessment center
- Other testing methods as deemed necessary
- Successful completion of field training officer school

Failure to complete or pass any portion of the selection/testing process immediately disqualifies an officer from further participation in the process. The officer must then wait for a period of one year from the termination date to be able to reapply.

Officers successfully completing all phases of the testing/selection process but not placed into active FTO status will be held in reserve status until being placed into active status or until the next application process is completed, whichever comes first. The training academy staff maintains an eligibility list of FTOs successfully completing the selection process and upon request, provides each service area major or deputy chief with the name of eligible officers in their service area to fill any vacant FTO position. As the need arises, officers on the eligibility list may be asked to serve in active FTO status to accommodate departmental training requirements. These officers are compensated as active FTOs from the time they are placed into active status until they are returned to the eligibility list. This temporary service is taken into consideration when additional officers are needed to fill active FTO status positions. Any officer on the eligibility list who at any time fails to meet any of the qualification criteria is dropped from reserve status and required to reapply when eligible.

A candidate must have been at least three years with the department and there is no maximum length of service to disqualify a candidate. After initial selection the candidate must complete a 40-hour training session at the Charlotte–Mecklenburg Police Academy. A 40-hour course of instruction was recently reinstated and replaced by the three-day course that has been offered. The content of this 40-hour course is compatible with the instruction presented in the previous blocks of instruction and is to be determined by the training academy staff.

The standards required of prospective FTOs are quite impressive and definitely an improvement over the minimum standards required of instructors in Virginia. The integrity-related components—the disqualifying factors related to the disciplinary record—are especially worthy of praise. However, the requirement to have a demonstrated commitment to the policy of community-oriented policing and problem solving poses a philosophical and very practical problem that, indeed, materialized in this department.

A number of problems can be identified when looking at this requirement:

1. If, from the philosophical standpoint, a person does not agree with the philosophy of community-oriented/problem-solving policing, he or she must hide these opinions to become an FTO.
2. If this is the case, the officer is forced to serve as a role model to a new recruit by hiding his or her real orientation.
3. Such an officer is not afforded the opportunity to be further educated in the concepts of the philosophy to which the department subscribes, since the officer has to keep silent about his or her views.
4. This, of course, causes further deterioration of an officer's morale and certainly does not generate the foundation for an effective FTO.

Since many officers do feel, for various reasons, that they want and can be good FTOs, they apply for the position and do not reveal their true feelings about the philosophy of community policing. The real-life scenario can then become extremely ugly, during an FTO meeting with a number of veteran FTOs complaining bitterly about the tenets of the philosophy and expressing their true feelings about certain concepts, which were very remote from those expressed in the mission statement of the department (Haberfeld, 1999).

The maturity theory can be applied once again to explain the foregoing phenomenon. The FTOs were obviously at level 3 readiness—able to perform as FTOs, based on their other qualifications as required by the department, but totally unwilling or insecure about the last requirement, commitment to the philosophy of community-oriented policing. One can easily picture the impact and effectiveness on new recruits of their instruction.

The voices of proponents of community-oriented philosophy can be heard claiming that somebody who does not subscribe fully to the philosophy of the department should not be an FTO. Two answers can be offered to this claim. First, since the philosophy of COP is interpreted in a different way by every police chief, it would perhaps be wise to define clearly what it is precisely that an officer has to subscribe to rather than simply referring to some elusive term. In addition, even when clear concepts are offered, not everyone has to agree that they can actually help "solve" problems related to crime and the community. The causes of a given crime are usually complicated and not easily changed by the tools and remedies that an officer can offer (even while working in close partnership with the community). Furthermore, not everybody is "born," or alternatively, "made" to be a problem solver—some officers will never display the desired level of readiness or maturity.

Second, one can be an effective instructor even if he or she does not prescribe to a given philosophy. One of the wonderful concepts in the world of education and training is the idea of different points of view. Of course, it is less practical in firearms instruction, but it certainly deserves some room in operational decisions based on philosophical concepts.

Incentives for Services as Field Training Officer

This part of the chapter deals with the incentives offered to FTOs and explains to a certain degree why officers might be tempted to lie about their true orientation toward community-oriented policing. The incentives offered are probably based on the assumption that good instructors can be "made."

The incentives provided to field training officers/evaluators for active service in a field training program are as follows:

- A pay increase of 3 percent is provided to officers serving on active FTO status.
- Each FTO/FTE may be assigned his or her own patrol vehicle for use on duty for the storage of training supplies at the discretion of the chain of command.
- Each FTO/FTE is issued an FTO pin which must be worn on his or her uniform while serving in the field training program.
- Each active FTO/FTE is among the first group of officers to be exposed to any new departmental training effort and may be asked to assist in development of new training.
- Each active FTO/FTE is among the first group of officers to be issued new equipment.

- Any FTO/FTE actively serving in the field training program is eligible to participate in the FTO/FTE of the year program.
- The FTO/FTE selected as the FTO of the year receives a departmental certification of acknowledgment.

This list of the incentives is pretty impressive and could serve as a real incentive even to an officer who knows that he or she is not ready for the task. It is extremely important to have the right incentives to attract the best possible pool of candidates, but it is even more important to make sure that the package being offered does not attract officers who are not ready to be FTOs.

Accountability of the Field Training Officer

Each FTO is held accountable for all training in which he or she participates or administers. In addition, each FTO is expected to serve willingly in any extra capacity deemed necessary within the chain of command and/or by the commander in charge of the training academy. All FTOs are required to attend periodic training updates as determined by the training academy staff. These training sessions are utilized to introduce FTOs to new field training methods or ideas and to present the opportunity for any FTO to submit any comments, suggestions, and/or problems to the training academy staff and to other FTOs for discussion. All FTOs are tested annually to ensure competency to continue serving in the field training program. Each FTO/FTE is held accountable for any and all training in which he or she participates or administers. This encompasses both training administered in the field training program and training provided in any other fashion within the department.

The presence of any of the following factors will disqualify any active FTO/FTE from participating in the field training program:

- Receiving below a "good" ranking on his or her annual PAR review
- Being rated poorly by probationary officers that he or she has trained
- Receiving low evaluations by his or her field training sergeant of participation in the field training program
- Violation of any of the original rules, standards, or qualifications of the field training program
- Any reluctance or unwillingness to participate or serve in the field training program when requested to do so by the chain of command or the training academy

Integrity-Related Functions of FTOs

The department has identified 31 essential job functions to be demonstrated effectively by field training officers. The integrity-related components are discussed in a more extensive manner since the CMPD appears to stress the importance of the records of disciplinary violations as a major disqualifier from becoming an FTO. Among these functions, the following stand out as being more closely related to the integrity components:

- Communicate effectively with people, including juveniles, by giving information and directions, mediating disputes, and advising of rights and processes
- Endure verbal and mental abuse when confronted with hostile views and opinions of suspects and other people encountered in an antagonistic environment
- Hold the police department—its policies, rules, regulations, general orders, goals, and objectives—in high esteem, as evidenced by speech, actions, attitudes, appearance, and overall personal demeanor
- Present a positive role model and a neat, professional appearance at all times
- Display self-motivation and initiative in teaching and training probationary officers in the field training program
- Display knowledge of the training materials utilized in the field training program and be able to perform adequately supervisory tasks needed to teach these police skills to probationary officers: serving as an instructor, teacher, trainer, and mentor
- Possess a positive attitude, desire, and motivation toward his or her work, his or her career, and the department
- Apply honesty, fairness, and objectivity in evaluating probationary officers on their performance during their field training

In addition to the emphasis on integrity-related qualifiers, the department has instituted two first-line supervisory levels: field training evaluator and field training sergeant. Both of these positions are concerned primarily with mentoring. Holding one of the two positions would require, at least in theory, additional mentoring skills, which, again, are not necessarily exhibited by senior-level trainers. (Further material on the level of readiness to serve in a supervisory position is presented in Chapter 8).

Field Training Evaluator

Each field training officer may be called to serve in the capacity of field training evaluator (FTE). The FTE evaluates each probationary officer for competency prior to the officer being released into solo status as a police officer. The FTE evaluates each probationary officer's overall performance and determines the competency of that officer in each training task. This final evaluation phase with an FTE consists of at least a 2-week period, with overall training and evaluation being conducted over a minimum of a 12-week period. Any probationary officer found to be deficient in any training task(s) is not allowed to advance to solo status until the deficiency is corrected. He or she will be placed into remedial training and an attempt will be made to correct those training deficiencies. This remedial training period should last a minimum of 4 weeks but no more than 8 weeks, with a review at the end of each 4-week period.

The FTE also evaluates the performance of each field training officer who has assisted in the training of each probationary officer in which the evaluator serves in the capacity of FTE. The evaluator presents the evaluation of each FTO to the field training sergeant for review and action, if necessary.

Field Training Sergeant

Each patrol district should have a designated field training sergeant whose responsibilities include the following:

- Monitor all district field training activities on a daily basis.
- Maintain complete and accurate records of all field training administered in his or her district.
- Evaluate each FTO/FTE.
- Conduct weekly and end-of-phase conferences with FTO/FTEs and probationary officers.
- Review all field training reports.
- Keep the training academy staff and his or her chain of command informed about any training deficiencies or problems.
- Ensure that each FTO/FTE remains eligible for participation as a trainer in the field training program.
- Ensure that active FTOs/FTEs are rotated for service in the field training program.

Each field training sergeant receives initial training on the field training program, its management, and on current training issues/methods administered to all FTOs/FTEs. Each field training sergeant should attend the same training updates as all active FTOs/FTEs. At the conclusion of each of these training updates, field training sergeants are afforded the opportunity to present any supervisory problems and/or any other ideas/comments/suggestions to the training academy staff outside the presence of the FTOs.

Midsized Municipal Agency: Charleston, South Carolina

The following information is relevant for departments that do not have their own in-house academies, and rely, in an even more profound manner, on the quality of their FTOs to introduce the recruit to the norms, values, and philosophy of their department. This particular agency sends its recruits to a state police academy, where they receive only minimal introduction to police work. The state academy in South Carolina offers instructional modules that cover only the most basic facets of law enforcement responsibilities over a (recently increased) period of 9½ weeks.

The vast burden of responsibility is therefore put upon the shoulders of the agency's FTOs. In theory the standards and qualifications to become an FTO are quite reasonable. In reality, however, and this is probably the case in most mid-sized and small police departments in the United States, these standards and qualifiers have to be compromised to accommodate personnel shortages. This is one of the primary reasons why we advocate the mandatory inclusion of the FTO training within the academy setting. As burdensome as it might look at first glance, this solution is the only real alternative the majority of police departments have at their disposal to ensure good-quality FTO training delivered by the most qualified, mature, and ready FTO officers.

The CMPD claims that their field training officers are carefully selected from police officers throughout the patrol forces. These officers are formally instructed in their responsibilities and duties as FTOs. The number of FTOs depends on departmental needs.

These officers should exhibit the following qualifications:

- Have exemplary reputations
- Demonstrate maturity
- Have a good knowledge of the penal code, traffic code, and departmental rules and regulations
- Be recommended by their supervisors
- Above all, exhibit a self-initiated interest in the assignment

They will be required to study the departmental field training manual and the recruit training manual and demonstrate a working knowledge of the principles outlined.

The selection process consists of the following chain of events:

1. A notice requesting volunteers is distributed to the entire police department. The notice lists minimum qualifications and due dates.
2. The training division completes a background report which includes information from the candidate's chain of command, internal affairs, training division, and other sources.
3. The training division obtains random samples of the officer's offense reports for the past several months.
4. Individual interviews are conducted (when deemed necessary) to clarify specific questions about the applicant.
5. The training division members meet and review all the information gathered.
6. The training division submits the names of acceptable candidates to the commander of patrol operations.

7. The training division and commander of patrol meet to discuss the candidates. A list of recommended candidates is prepared at this time.
8. The commander of patrol operations submits the list of recommended candidates to the chief of police for final approval.
9. All candidates are notified, in writing, of the results of the selection process.
10. Immediately upon selection, FTOs are issued a copy of the FTO manual and begin to familiarize themselves with its content. As soon as possible, the new FTO is taught the approved 40-hour course provided by the state academy.

Beyond the fact that the selection process addresses some critical issues that other departments ignore, such as exemplary reputation and maturity, the mission of selecting and preparing the best FTO is not completed. The problem lies in the scarcity of available personnel. It is one thing to outline the necessary steps but it is another entirely to be able to implement the plan. From a number of interviews with the FTOs in that agency (Haberfeld, 1999), a grim picture was provided. It is not an issue of "made," "born," or "mature" individuals, or even the most enthusiastic on the force. Due to the shortage of personnel, officers barely out of the probationary period (one actually still in the final week of the probationary period) became FTOs. It is, indeed, not difficult to imagine what kind of insight into police work such a "green" FTO can offer. On the other end of the spectrum, a veteran officer with 17 years on the force also serves as an FTO. This is not to say that a veteran of 17 years of police work cannot be a good FTO, but police work is "high-voltage" work. Seventeen years of service in any department will take its toll on an officer. Cynicism, or as the author prefers to refer to it, "moral corruption," developed through exposure to the well-documented stressors that police officers encounter during their career, cannot be ignored. These stressors can definitely make one stronger, wiser, and experienced, therefore contributing to a person's overall maturity. However, they can also produce a negative role model, and the agency should be extra careful in its selection process when dealing with veteran FTOs, who are products of traditional training. (For additional analysis of the negative influence of years of service on morale, see Chapter 7).

Duties and Responsibilities of the Field Training Officer

The FTO has complete responsibility for recruits during the field training assignment. The recruit always works with the assigned FTO and has the same work schedule. The FTO has two primary roles to fulfill:

1. As a police officer assuming full beat responsibility
2. As a trainer and role model for the probationary officer

The FTO must have the requisite skills to become a reliable evaluator of a probationary officer's performance. He or she is required to write evaluations of this performance and submit the additional documentation required. The

FTO must ensure that the recruit be familiarized with headquarters operation, including communications operations, record facilities, central Investigative division, and evidence/property room. He or she is the essential means by which the goals of the program are achieved, specifically the production of a police officer able to work a solo assignment in a safe, skillful, productive, and professional manner.

To achieve the foregoing qualifications, the FTO's performance is also subject to evaluation: His or her professional and personal conduct should be exemplary. The FTO should possess, and recognize the need to possess, even a higher sense of idealism than that generally found throughout the police department. The importance of maintaining a student–teacher relationship cannot be overemphasized, and the FTO is also responsible for ensuring that the probationary officer will write a brief statement at the end of each phase detailing his or her own strengths and weaknesses in the training program. Finally, the FTO is charged with the responsibility for recommending termination of a probationary officer when retention is no longer productive.

FTOs receive some assistance for the evaluation process, termed standard evaluation guidelines. The 7-point rating scale discussed earlier serves as a basis for evaluating and rating a probationary officer's performance. As guidelines, these definitions serve as a means of program standardization and continuity. Following are a number of examples that might be problematic in an adequate evaluation.

Appearance

1. *General appearance:* Weight must be appropriate for height and build. Uniform and equipment must be appropriate for varying situations and must be clean and neat. *Hygiene:* Offensive body odor and/or bad breath are not acceptable. Hair and mustache must be neatly trimmed to conform to regulations.

    (1) *Unacceptable:* overweight; dirty shoes, equipment, or uniform; long, unkept hair; offensive body odor.

    (4) *Acceptable:* neat, clean uniform and weapon; well-groomed hair; shined shoes.

    (7) *Superior:* tailored, clean, and pressed uniform; "spit-shined" shoes and leather.

It is interesting to observe again that weight serves as a disqualifier in police work. What if the FTO himself or herself is overweight? Does the agency send a message here—that a person's weight can be controlled by the person? That someone who is overweight cannot be a good police officer? If yes, what kind of message is the FTO conveying to the officer? The implications are left to your imagination.

Attitude

2. *Acceptance of criticism (feedback):* Is corrective criticism accepted in the manner given? Is there resentment as a result of the criticism? Is the behavior following corrective criticism positive? Was the probationary officer able to turn the criticism into productive behavior?

   (1) *Unacceptable:* rationalizing; argumentative; fails to make corrections; considers criticism negative.

   (4) *Acceptable:* accepts criticism in a positive manner and applies it to further learning processes.

   (7) *Superior:* solicits criticism in order to improve performance; never argues or blames others.

The superior attitude is worth further comment. The FTO is supposed to prepare the officer to be a problem solver. To be an effective problem solver, people need to challenge certain, existing situations, conditions, and circumstances. To solve a problem, one sometimes needs to argue. Is a given FTO skilled enough to make a true distinction between a person who is argumentative just for the sake of it and somebody who critically approaches the situation and looks for the best possible solution? In a college environment. students are constantly encouraged (or should be) to think critically. It is sometimes difficult, even for a college instructor, to distinguish between a genuine thinker and a troublemaker. Perhaps the assessment techniques required of FTOs in Charleston should be reassessed.

Attitude toward Police Work

3. From what the probationary officer says and does, is his or her attitude toward police work positive?

   (1) *Unacceptable:* takes police work as only a job; uses job for ego trips; abuses his or her authority; no dedication; fails to utilize off-duty time to further professional knowledge.

   (4) *Acceptable:* expresses active interest toward the job.

   (7) *Superior:* utilizes off-duty time to further professional knowledge; maintains high ideals toward professional knowledge; maintains high ideals toward professional responsibilities.

Field Performance: Stress Situations

18. Under conditions of stress, how acceptable is the probationary officer's field performance?

    (1) *Unacceptable:* becomes emotional and panic stricken; unable to function; loses temper; takes improper or no action.

(4) *Acceptable:* exhibits calm and controlled attitude; does not allow his or her own actions to further aggravate the situation; able to assess the situation and take proper action.

(7) *Superior:* maintains control and brings order under any circumstance without assistance.

Is it reasonable to maintain control and bring order under any circumstances without assistance? One of the desirable skills of the ideal instructor is a sense of humor. Indeed, an FTO will have to exhibit a lot of this if he or she is looking for superior probationary officers: the superman or superwoman among the trainees, a person who, in addition to being in total control under any circumstances, fully utilizes off-duty time to further professional knowledge. It is very important for an instructor's morale (be it a trainer or an educator) to be in a situation in which he or she can grade a student as superior. Standards for such "superiority" have to be realistic; otherwise, the instructor will end up having only mediocre performance from the probationary officer, which will ultimately result in negative self-perception and a negative evaluation form filled out by the probationary officer in assessing the FTO.

Control of Conflict: Voice Command

22. In conflict or potential conflict situations where voice commands are given, were the results positive?

(1) *Unacceptable:* improper voice inflection (i.e., too soft, too loud); confused voice; indecisive voice command; poor officer bearing.

(4) *Acceptable:* speaks with authority in a calm, clear voice.

(7) *Superior:* gives appearance of complete command through voice tone and bearing.

Improper voice inflection is a speech impediment that can easily be corrected. An FTO instructor should not be asked to assign an unacceptable rating regarding a disability that can easily be corrected. Such a rating sends a demoralizing message to a probationary officer.

Use of Common Sense and Good Judgment

23. To what degree of acceptability does the probationary officer possess and employ common sense and good judgment in police situations?

(1) *Unacceptable:* acts without thought or is indecisive or naïve.

(4) *Acceptable:* able to reason out a problem and relate it to what he or she is taught; good perception and ability to make his or her own decisions and carry them out.

(7) *Superior:* excellent perception in foreseeing problems and arriving at correct solutions in advance.

If we were only able to foresee problems and arrive at "correct" solutions in advance, this world would be so much nicer! We might not even need a police force. Again, the author leaves it to the imagination of the reader and the course instructor to discuss each of the points presented in this chapter. This reality check is strongly recommended, as a class exercise. Nothing demoralizes an instructor, educator, or student more than unattainable performance goals.

Small Municipal Agency: Washington Township, New Jersey

The most indicative of the problems in selecting ideal instructors or educators can probably be found in small police departments across the United States (50 or fewer sworn officers), which constitute over 95 percent of the police agencies in this country. The author once attended a professional conference during which a paper entitled "Small Police Departments—Let's Just Leave Them Alone" was presented. It is possible that in certain respects, the author of this paper could present some arguments that would justify the title. However, in reference to training, one should probably raise a hue and cry while disagreeing completely with the premise of "leaving them alone."

Small police departments deserve all the help and assistance they can get in identifying, selecting, and making full use of officers who are ideal instructors. Unfortunately, the way training is structured today, leaving FTO phases outside academy training does not afford the best possibilities.

Selection Process for FTOs

1. Police department posts notice of field training openings.
2. Written application has to be filled by a prospective FTO.

Minimum Qualifications to Become an FTO

1. Five years of employment as a sworn officer with the police department, of which six of the 12 months prior to the FTO application date were served in uniformed patrol service
2. Acceptable departmental performance evaluations

In addition, applicants shall be assessed by a selection board composed of four members, one of whom is an active FTO. The FTO candidate must attain a minimum score of 75 percent, or 31.5, on the field training officer candidate evaluation. Final field training officer selection shall be made by the chief of police and/or designee. To be certified, all FTOs must attend a 40-hour field training officer class. This may be waived, temporarily, by the chief of police. It is left to the imagination and the simple mathematical calculation of the reader how selective a department with 30 officers (including the ranking officers) can be. The fact that the chief can temporarily waive the requirement to attend the FTO class speaks for itself. The "abundance" of minimum requirements to become an FTO in this agency completes the picture. Unfortunately, this small police department is representative of the problems that other small agencies face, and in its own way is in a much better situation than most of the others (a very thorough and well-prepared FTO manual and evaluation program).

STRATEGIES FOR TRAINERS

It is beyond the scope of this chapter to tackle all the critical issues embodied in the process of selection of instructors and trainers. It is sufficient to mention a couple of major deficiencies and point to an alternative direction; the rest can be left to the creativity and resourcefulness in problem solving of each police department. The final part of this chapter is devoted to the strategies identified by the British Police Training Center and the recommendations of the author. While revising its training modules in 1999, the British Police Training Center identified a number of strategies that are potentially useful for police trainers. These strategies are based on the analysis of the natural processes of adult learning.

1. If learning is episodic, the best effects could be achieved by breaking up material into manageable units. Each unit needs to be linked to other items of learning and the goals and subgoals.
2. If learning is focused on concrete and immediate goals, the material must be relevant to the present needs of the learner. They should be encouraged to set short-term goals themselves.
3. If learners are to be motivated, relevant issues must be tackled immediately. The beginning of the manual is not necessarily the best place to start.
4. Learning by doing a task is more appropriate to adult learning than just preparing to be able to do it.
5. Where learners have their own patterns of learning, trainers should discover what they are and devise activities that capitalize on them and develop them further.
6. Trainers should not do things for adult learners; both sets of people should participate in the teaching–learning process.
7. Trainers should encourage adult learners to relate new material to their own experience where possible and make that experience available to the entire group.
8. The use of trial and error should be encouraged by participatory and discovery learning methods. Learners need to be active, not passive, participants of information and instruction. There is a need for practice and reinforcement. Constant feedback should be built in to programs. This takes time. Adult learners need time for learning.

9. It is just as necessary for adult learners as for younger learners to move from simple processes to complex ones. Learners should be helped to build up units of new material to create what they require.

10. The decline of memory in adults indicates that more reliance should be placed on understanding rather than retention. Memorization or role learning on its own is inappropriate.

11. Demonstration is an essential part of helping adults to learn. Participants should not normally be expected to perform a task without first having worked through an example with the group.

12. Trainers should start with concrete issues and move to more general issues rather than the other way around. The use of particular instances and their implications should be used to arrive at general conclusions, and relationships should be build up between different units of material used in learning.

13. Adult learners should be encouraged to continue the process of learning outside the classroom. Learners should be encouraged to learn without the help of trainers. The trainer's main aim should be to create effective learners.

Although the ideas identified by the British police are definitely worth emulation, following are some more modest recommendations, less conceptual in nature but easier to consider and implement.

1. Identify the minimum qualifications based on the topic or/and nature of instruction.

2. Match the minimum qualifications with objective professional standards and do not try to customize and compromise the standards based on a given philosophy. It is beneficial to expose trainees to different points of view as long as somebody oversees the overall outcome.

3. Expand your instructor resources by including the ones without the "obvious" minimum qualifications. For example, an effective diversity trainer does not necessarily have to be a high school graduate.

4. Look for the "new skills," such as sense of humor. Critical thinking skills can be found in somebody who is argumentative—it might mean that the proverbial troublemaker might actually be a better instructor than the rigid "yes man" rule follower.

5. Balance the "new skills" and "traditional skills" instructors. Police work entails emergency response situations during which there is no room or time for arguing or for too many of the new skills. There are situations in which military drill type of instruction based on the philosophy "stop thinking and start following the orders" is the only way to act.

6. The skills of advanced and specialized trainers are discussed in Chapter 11. However, everything that has been depicted as problematic in reference to the academy and FTO trainers also holds true for specialists. To hire a college professor to deliver a module on community-oriented policing simply

because he or she is a college professor misses the point of the instruction. A college professor who has a very limited (if any) knowledge of real street police work (and the nature of an interaction between a police officer and a citizen) will deliver a very ineffective training module. It is imperative to select a specialist based on a much broader consideration other than his or her current work affiliation. (For further discussion of this topic, see Chapter 9).

7. Finally, police departments that do not have an in-house academy or the necessary personnel to provide the best possible instructors might designate one or two who are the best and rotate them with neighbors. In emergency situations, small police departments frequently rely on a backup system provided by the neighboring forces. Why not rely on neighbors to provide additional FTOs? The argument against this system can already be heard: "We have our own standards and they do not measure up." The answer would be simple: Find those who do measure up; it is certainly a better solution than employing officers barely out of their probationary period as instructors and role models for the new recruits.

SUMMARY

The list of critical issues in police training becomes a few pages longer with the problems identified in the chapter. Police organizations are by no means in a unique position in their quest for quality manpower to prepare their future and current employees for the hurdles of their profession. However, while other professional organizations, schools, colleges, and universities, strive to identify and recruit the best their money can buy, police organizations, unfortunately, are nowhere near a similar approach.

The chapter outlined a number of problems with which agencies struggle, on a daily basis, while trying to identify the qualified instructors and educators. While it is hard to always attract the best and the most qualified, the problem seems to be further emphasized by the misguided approach to the selection process. The candidates are frequently chosen based on the available rather than desired, and further on screened as to their suitability based on the their physical condition and overall appearance rather than more learning and teaching-oriented qualifications.

The chapter culminates with a list of recommendations, aimed to assist the people who are responsible for the hiring and retention of police trainers and educators.

REFERENCES

Argyris, C. (1957). *Personality and Organization: The Conflict between the System and the Individual*. New York: Harper.

Bakke, E. W. (1950). *Bonds of Organization*. New York: Harper.

Berliner, D. C. (1986). In pursuit of the expert pedagogue. *Educational Researcher, 15*, 5–13.

Biehler, R. F., and Snowman, J. (1993). *Psychology Applied to Teaching*, 7th ed. Boston: Houghton Mifflin.

British Police Training Center (1999). *Probationer Training Manual*.

Charleston, S.C. Police Department (1998). *Field Training Manual*.

Charlotte–Mecklenburg Police Academy (1999). *F.T.O. Training Manual*.

Haberfeld, M. (1999). Field notes.

Hersey, P., and Blanchard, K. H. (1977). *Management of Organizational Behavior: Utilizing Human Resources*, 3rd ed. Englewood Cliffs, NJ: Prentice Hall.

Hersey, P., and Blanchard, K. H. (1993). *Management of Organizational Behavior: Utilizing Human Resources*, 6th ed. Englewood Cliffs, NJ: Prentice-Hall.

Highet, G. (1957). *The Art of Teaching*. New York: Vintage Books.

Klotter, J. C. (1978). *Techniques for Police Instructors*, 4th ed. Springfield, IL: Charles C Thomas.

Police Training Council. (1999). *PTP Training Manual*. Bramshill College, England.

Shulman, L. S. (1986). Those who understand: knowledge growth in teaching. *Educational Research, 15*, 4–21.

Washington Township, New Jersey, Police Department (1999). *Field Training Manual*.

STRESS MANAGEMENT TRAINING
Feelings–Inputs–Tactics

INTRODUCTION

This chapter is the first of the four "practical" chapters, which in addition to the theoretical perspective and descriptive depiction of the state of a specific straining situation in a given agency, offer a very pragmatic approach to a certain module of training, in this case, stress management training. The practical chapters offer a limited overview of the currently utilized modes of training in a given area but, instead, concentrate on a new and desirable approach to the specific topic. Each of the four chapters differs in the extent to which the emphasis of the chapter is placed on the theory behind the proposed module, the descriptive part of the training currently offered, or the newly designed module.

This chapter was developed by the author for an FBI-sponsored conference, "Suicide and Law Enforcement," which took place in Quantico, Virginia, in September 1999. Some of the information presented in this chapter is already irrelevant. For example, the outcome of the Diallo shooting case is already known, and the officers were actually acquitted. Nevertheless, to fully understand the premise on which the proposed training module was developed, the author decided not to change the initial submission. The situation depicted in the "41 shots" scenario, despite being very specific, illustrates the overall struggle of police officers accused of premeditated actions. It can easily be replaced, almost on a daily basis, by situations in which officers are accused of violating the human rights of the people they police.

As this chapter was being written, a number of police officers from the Philadelphia Police Department were being portrayed in the national news media abusing the rights of their office by kicking and beating a car-jacking suspect. However, the author and other "spectators" have a very limited knowledge of what actually happened on the streets of Philadelphia that caused the officers to react the way they did. Nevertheless, limited knowledge has not prevented numerous media "crusaders" from publishing inflammatory editorials on the pages of our daily newspapers.

In this chapter we attempt to put stress management training in the right perspective, that which will enable the officers to deal effectively with the presumption of "guilty until proven innocent," and even when proven innocent, by a jury of their peers, still perceived as being guilty.

THE STARTING POINT

Police work is a misunderstood phenomenon. People tend to romanticize, stigmatize, demonize, exaggerate, and mostly misunderstand the critical aspects of police work. It is not about danger, power, esteem, or politics. It is first and foremost about a very special calling, which enables one person to sacrifice his or her own safety and security in order to protect others. It is about priorities set by police officers, which are almost antithetical to common sense, in which a person puts other people's needs ahead of his or her own. However, being able to elevate oneself above and beyond common sense is not without a price: recognition of the sacrifice. If, however, recognition is missing, factually or perceptionally, a police officer may embark on a profoundly ruinous road, toward cynicism and self-destruction. We begin with a depiction of the poorly understood aspects of police work. Three real-life encounters involving frustrated and misunderstood police officers highlight the dire need for an expanded definition of critical incident stress. The expanded definition is followed by revisiting and extending Maslow's hierarchy of needs, which, in turn, provide fertile ground for new training modules. The new concepts are presented in a generic mode, and the author recognizes the need for customization, based on the size and the resources of each and every department.

The CompStat Meeting

It is an early morning hour in the command and control room of the New York City Police Department. The room is already partially filled by guests of the police commissioner and some officers. At 7 A.M. sharp, the meeting starts. The room, although filled primarily with law enforcement personnel, seems to be divided by an invisible line: on one side the departmental brass, who will lead the meeting, and on the other side those who are going to respond. The meeting lasts for three hours, during which a number of officers, from high-ranking precinct commanders to plain clothes detectives, answer a battery of aggressive questions directed at them. The entire encounter resembles a high-intensity football game more than a departmental meeting. The big screens behind the backs of the "defensive team" light up with numbers and statistics, adding to the overall atmosphere of an offensive attack, far from what one can define as constructive criticism. It feels like the tension in the room could be cut with a knife. The team on the defense holds up quite well; however, from time to time one could see a dangerous spark in the officers' eyes. As accusations fly, from the perfectly valid to the little bit extreme, a short break is announced. In a way, this brief intermission might be considered as a time for regrouping, for the team on the offense, since the moment the meeting begins, the vitality of the inquisitors seems more powerful than before. The questioning continues, and the officers bravely face the mounting attacks. Finally, three hours later, at 11 A.M., the meeting is over. No blood was spilled, nobody got hurt physically, the overall productivity or clearance rate of the department might even go up; however, no one has addressed the psychological impact. "It works" are the words frequently spoken by the departmental brass. Yes, maybe it does, technically speaking, for a relatively short period of time, but for an outsider sitting in this room, charged with high-intensity verbal assaults, assaults flying in one direction only, "it doesn't work." One can only speculate what it would take for an outsider to endure this treatment. In a football game the players often get hurt, generally physically, but mentally as well; that is the price they have to pay for fame, money, recognition, and adoration. After the CompStat game is over, the police officers go back to the streets or their precinct commands. There is no fame, no money, no adoration, no recognition, and most of all, often there is no justice for them. It goes beyond saying that in any work environment there is always room for improvement and accountability. One question remains unanswered, though: What is the right way to express constructive criticism? Is the bottom line the final technical outcome, or is the bottom line human dignity? One could argue that the two are equally important: the clearance rate and/or reduction in crime and the officer's morale. However, it is clear that only one variable of this equation is taken into consideration during a CompStat meeting.

The "41 Shots"

It is reasonable to assume that most readers are familiar with the tragic events that led to the death of Amadou Diallo in New York City. Four plain clothes police officers, members of the NYPD's Street Crime Unit, received some intelligence information about a rape suspect. Following the intelligence lead, the four entered into an encounter with Amadou Diallo, an African immigrant who came to the United States to improve his living conditions. What exactly happened during the encounter is subject to speculation but the fatal outcome is a fact. This case of mistaken identity led to 41 shots being fired at the unarmed Diallo, resulting in his immediate death. The officers involved were charged with second-degree murder.

It is beyond our scope here to provide a defense of the four officers, a case that might be stated quite legitimately while taking into consideration a person's physiological state during critical incident stress, one of whose symptoms is a significant impairment of peripheral vision: up to 70 percent may be impaired by stress (Olson, 1998). This symptom may serve as a valid explanation of why the officers fired that many shots and of why they hit the victim in his legs and other parts of the body (impaired vision causes a person to shot at the wrong target). Regardless, the crucial point is not a defense of the officers but the offense they were charged with. Second-degree murder implies that the officers, although without premeditation, intended to kill Mr. Diallo.

Despite the fact that this incident is a perfect example of critical incident stress, from the perspective of the accused officers, the core of this incident lies not only in the fact that the life of an innocent man was taken away but that they are accused of taking away his life intentionally. One cannot begin to imagine how it must feel to make a tragic mistake in the course of one's line of duty and not only suffer the consequences of this mistake but be exposed to charges of intent. No matter what the final outcome of this case, for the defendants, there is no justice for them as much as there is no justice for any officer on the street who attempts to serve and protect the society from rapists, murderers, and other dangerous persons. "Mistakes are human" people frequently say, but it seems that police officers are excluded from this categorization. Soldiers occasionally get killed by friendly fire—it's tragic and inexcusable, but it happens. Very infrequently, if ever, are the soldiers involved charged with a second-degree murder.

The Off-Duty Encounter

A police officer in his early twenties is having a beer after work. After one beer the officer leaves the bar and approaches his car. The car is blocked by a double-parked vehicle. The officer approaches the driver and asks him to move the car. The driver refuses and curses the officer, who at this point identifies himself as a police officer (he is not wearing a uniform). The driver looks at the officer's ID and pulls out a gun. At this point the officer pulls out his weapon, and at the same time a police car arrives at the scene, followed by another patrol car, and the arriving officers take control of the situation. The citizen is handcuffed and taken away and the young officer is asked for his statement. While the statement is been taken from the officer, his gun is taken away from him by the officer in charge. The next day he is placed on suspension without pay for drinking and displaying his weapon off duty.

This story was told to the author by one of her students, who happened to be the young officer depicted above. The young man, now with three years on the force, was taking the class "Police and Community Relations" and decided to share his personal experience with his classmates during a class entitled "The Human Experience of Being a Police Officer."

It is beyond our scope here to analyze the truthfulness of the story; nevertheless, it must have some validity, as toward the end of the semester the officer was reinstated and returned to his regular duties without a disciplinary hearing, according to his report. During the semester he came to the author a number of times to express his frustration with the system—the police department that doubted his words and violated his trust, trust based on the assumption that in a hostile encounter with a citizen (on or off duty) he would receive backup from the organization. The importance of this story lies not so much in its accuracy but in the fact that the officer, loaded with frustration, obviously had no outlet for his grief. The author asked him a number of times, and so did other students in class, whether he had complained, received counseling, and so on. The answer was cynical: "No, who cares; the organization doesn't care about you, and your colleagues are too preoccupied with their own stuff."

The story demonstrates that on a daily basis, officers emerge from encounters with citizens, peers and supervisors with a feeling that the organization and/or their peers did not provide them with the support they felt they were entitled to. Although an incident may not seem critical, the sense of injustice can be very real.

Police Suicide as a Function of the Routinely Ignored Hidden Stressors

The desired response to critical incident stress would be counseling, debriefing, and so on, but even if provided, this range of responses would address the wrong stressors. The stressors addressed would, undoubtedly, include the tragic situation itself, the death of an innocent person, and the entire killing situation. It is, however, doubtful that the profound injustice embedded in a charge of second-degree murder would be addressed immediately or ever, by the police organization. After all, officers involved in cases such as those above are generally placed on suspension; therefore, by default they are guilty until proven innocent and there is no room for debriefing on the charges. Whether they are found guilty or innocent of the charges, it is extremely improbable that they will receive counseling or other assistance. If they are found guilty, there they go, and if they are found innocent, they will be reinstated, but once again the hidden stressors will be ignored. One is expected to deal with these stressors and even accept them as justified. (As one of the author's colleagues said, precinct commanders are well paid and should be accountable for their work, whatever it takes.)

Our point here is that incidents can generate critical incident stress disorders which if not recognized and treated can lead to cynicism, depression, and in extreme cases, to suicide. Some answers to the problem of police suicide stem from the ill-defined and misunderstood phenomenon of how police officers react to situations in which they believe that injustice is being imposed on them.

STRESS MANAGEMENT TRAINING IN LAW ENFORCEMENT

Critical Incident Stress Definition as the Source of a Misguided Approach to Training

In the past, most studies of stress in law enforcement focused exclusively on postshooting trauma. Kureczka (1996) identified a number of other traumatic events, collectively known as critical incident stress (CIS). His definition encompasses any event that has a stressful impact sufficient to overwhelm a person's usually effective coping skills. Among the events listed are a line-of-duty death or serious injury of a co-worker, a police suicide, an officer involved in a shooting situation, a life-threatening assault on an officer, a death or serious injury caused by an officer, an incident involving multiple deaths, a traumatic death of a child, a barricaded suspect/hostage situation, a highly profiled media event, or any other incident that appears critical or questionable.

According to Kureczka, the definition of a critical incident must remain fluid because what affects one officer might not affect another. This particular assumption is extremely valid for the expanded definition of the CIS that will be presented later. In 1980, the American Psychiatric Association formally recognized the existence of a disorder similar to that referred to by the military as battle fatigue, which became known as *posttraumatic stress disorder* (PTSD). Symptoms of the disorder include intrusive recollections, excessive stress arousal, withdrawal, numbing, and depression. Pierson (1989) claims that critical stress affects up to 87 percent of all emergency service workers at least once in their careers. Critical incident stress manifests itself physically, cognitively, and emotionally.

Walker (1990) provides a slightly different definition of a critical incident. He describes it as ". . . any crisis situation that causes emergency personnel, family members, or bystanders to respond with immediate or delayed stress-altered physical, mental, emotional, psychological, or social coping mechanisms." She recognizes the need for critical incident stress debriefing procedures, using Mitchell's (1983) process, which includes the elements of factual description of the event, emotional ventilation, and identification of stress response symptoms.

Stress Management Training as a Function of the Ill-Defined Problem

The approaches to the CIS described above are among those common to the problem, and stress management training modules devised by and for various law enforcement training academies rely heavily on those definitions. Finn and Esselman Tomz (1997) published a thorough manual about developing law enforcement stress programs, but this publication seems to suffer from a similar disease—multiple and intangible definitions. Overreliance on fluid and elusive terms on one hand, and on an infinite host of traditional traumatic events (e.g., shootings, deaths, injuries, etc.) on the other, provides for a misguided approach to training.

The problems enveloped in CIS are ill defined and inadequate. First, one cannot devise an effective training module if one cannot define, and define precisely, what it is that you would like your recruits to be trained in, against, for, and so on. Undoubtedly there are a number of good definitions offered by researchers, but those definitions cover only a small percentage of the problematic issues involved in critical incident stress. If as the researchers claim, the definition must remain fluid, since what constitutes a critical incident for one officer might not affect another, the only rational conclusion is that we must abandon stress management training since we are targeting only a very small percentage of our audience. It is extremely difficult to identify with certain situations that are supposed to generate certain feelings and emotions, when one cannot generate those feelings since the situations presented are not relevant to one's emotional buildup. In a given training environment, the theoretical depiction of events, no matter how realistic and potent, remains theoretical for a significant segment of the audience. The examples cited by researchers, such as the death of a partner, death of a child, or traumatic media event, remain in the sphere of "unreal", since the training is offered to recruits who still do not have a partner, most of whom do not have a child, and who cannot possibly envision the power and influence of the media on their daily performance. When stress management training is offered only to the officers who are already on the force, new recruits enter the workforce exposed to the dangers of being affected by the CIS and having no coping mechanism whatsoever, or the ability to recognize the danger.

To emphasize how important an effective training module to a definition of a problem is, one might want to examine a number of traditional training topics, such as stress during night fire (training module offered by the New Orleans Police Department). It is impossible to envision this training module being offered to anybody without a clear definition of the problem, which would probably include the fact that this stress could only be developed under night-light circumstances. If on the other hand, this particular module would start with a fluid and elusive definition (e.g., you might encounter this stress during night shooting and perhaps also in other circumstances), the effectiveness of the module becomes highly questionable. Therefore, when offered to law enforcement officers, current stress management training is clearly a product of an ill-defined problem.

Redefining Critical Incident Stress

"It's a cop thing. You wouldn't understand." The new and expanded definition of critical incident stress offered here is based on the assumption that police officers, en masse, join law enforcement agencies to serve and protect the public from "bad guys." These sentiments were adequately defined by researchers. Crank (1998) believes that police see themselves as representatives of a higher morality embodied in a blend of American traditionalism, patriotism, and religion. According to Sykes (1986), as moral agents, police view themselves as guardians whose responsibility is not simply to make arrests but to roust out society's troublemakers. They perceive themselves to be a superior class (Hunt and Magenau, 1993) or as people on the side of the angels—the sense of "us versus them" that develops between cops and the outside world forges a bond between cops whose strength is fabled (Bouza, 1990). Police believe themselves to be a distinct occupational group, apart from society (Van Maanen, 1974). This belief stems from their perception that their relationship with the public, with brass, and with the courts is less than friendly, sometimes adversarial.

As outsiders, officers tend to develop a "we–them" attitude, in which the enemy of the police tends to shift from the criminal element to the general public (Sherman, 1982).

Police are haunted by accountability. They are in an occupation where situations in which they intervene are unpredictable, and sometimes they have to make rapid-fire judgments in emotional circumstances. Cops know that they will make many mistakes for which they would be publicly rebuked by any of a number of groups, the press, civic organizations, and departmental brass. Each of these is an influential actor in the cop's world and career (Crank, 1998). Furthermore, they joined the force to serve and protect all those influential actors, who so frequently, are scrutinizing their performance.

"To serve and protect" entails, at least in an officer's mind, delivering justice. In other words, the "good guys" (the police officers) are here to enable "us" (members of the society) to live in a civilized manner, protected, or at least under our Constitution being entitled to protection from the "bad guys." This profound, sometimes subconscious belief enables us to function on a daily basis without looking over our shoulders for predators and enemies. This sense of security is almost built into our civilized systems—we "know" that around us there is an invisible fence of protection provided by law enforcement officers. Of course, sometimes we experience some erosion in our sense of built-in security, predominantly when we are involved in an incident from which we emerge injured—physically, psychologically, or both—since there was nobody out there to protect us on an immediate basis. This sense of insecurity could be extremely traumatic for the rest of a person's life, and frequently, one cannot regain the built-in feeling of security.

Despite serving as protectors from evil and messengers of justice, no matter how symbolic, police officers have the same built-in need for security and justice, even though they are supposed to provide these needs for themselves. They are fully prepared, at least mentally, to do so, but as opposed to a citizen, they often face the reality of danger and injustice. From this assumption a new and expanded definition of CIS is presented: Critical stress incident can be generated by any situation or encounter with a citizen, peer, organization, or other from which a police officer emerges with a feeling or perception that justice has not been served for him and/or others.

The sense of being on the right side, on the side of the angels, crumbles when an officer realizes that although he or she is expected to provide justice for others (again in a symbolic way by serving and protecting the good citizens from the bad ones), there is no justice for him or her. The built-in mechanism that produces the faulty but effective sense of safety and security disintegrates and the sense of fairness disappears, leaving a residue of fear and cynicism, a proven formula for stress. Each of the three incidents described at the beginning of this chapter could be defined as CIS, based on the aforementioned definition. The frequent accumulation of such encounters, which seem to be present routinely in police work, is conducive to depression, mental breakdown, and in the most extreme cases, police suicide.

Maslow's Need Hierarchy Revisited

The Missing Link in Maslow's Need Hierarchy

Probably one of the most popular motivational theories is the one developed by Maslow (1954). He postulated that people's needs were exceedingly complex and were arranged in a hierarchy. His theory of motivation is based on the assumption that human beings are motivated by a number of basic needs that are clearly identifiable as species-wide, unchanged, and instinctual. This theory identified five need categories:

1. *Physiological needs.* These are the strongest and most fundamental needs that sustain life, and include food, shelter, sex, air, water, and sleep.
2. *Security needs.* These needs emerge once the basic needs are fulfilled. The dominant security needs are primarily the need for reasonable order and stability, and freedom from being anxious and insecure.
3. *Social (or the original belongingness and love) needs.* With fulfillment of the physiological and security needs, the social needs emerge. Human beings will strive for affiliation with others, for a place in a group, and will attempt to achieve this goal with a great deal of intensity.
4. *Esteem needs.* There are two categories of esteem needs. The first is self-esteem, including such factors as the need for independence, freedom, confidence, and achievement. The second is respect from others and includes the concepts of recognition, prestige, acceptance, status, and reputation.
5. *Self-actualization.* When most of the esteem needs are fulfilled, "what man can be, he must be." This is the stage of self-actualization characterized by the need to develop feelings of growth and maturity, to become increasingly competent, and to gain mastery over situations. Motivation is totally internalized; external stimulation is unnecessary.

Maslow did not view the hierarchy of need as a series of totally independent levels. In fact, the categories overlap and are not entirely precise. He suggested that the unsatisfied needs influence people's behavior. Since the initial research, Maslow (1970) developed a new list of needs, identified as growth needs (social, self-esteem, and self-actualization) compared to basic needs (physiological and safety). The higher needs, or growth needs, utilize the basic needs as a foundation. These higher needs are wholeness, perfection, completion, justice, aliveness, richness, simplicity, beauty, goodness, uniqueness, effortlessness, playfulness, truth, and self-sufficiency. (The need for justice can be related to the sense of injustice discussed above.)

These values are interrelated and cannot be separated. One should not make the mistake of thinking that satisfying one need, such as a good salary, will automatically transform all employees into growing, self-actualized persons. When the needs are not fulfilled, the lack of satisfaction generates certain behavioral patterns.

Unfulfilled physiological needs can generate pain, suffering, possible impairment, discomfort, or illness. Unsatisfied security needs might cause stress, anxiety, fearfulness, or trepidation. Feelings of being alone, remote, sad, or unloved can be caused by a lack of social needs. Insecurity, lack of a firm belief in one's own power, may be a result of unfulfilled self-esteem needs. Finally, when self-actualization needs are missing, the result is alienation, bitterness, frustration, or a feeling of uselessness.

What appears to be missing from Maslow's typology is one basic need that could probably be included in the category of basic needs—physiological security (or safety)—the need to communicate, or to put it in plainer terms, the need to speak one's mind.

As much as a human being needs food, shelter, sex, air, water, and sleep, he or she also needs to express his or her unique thought processes. From earliest times, as evidenced in the most ancient archeological sites, it is apparent that even though most effort had to be devoted to the satisfaction of basic needs, people found time to paint and draw, the first primitive form of expression of their thought processes. The need to speak or to communicate enables people to move on to the next stage of Maslow's hierarchy—to fulfill their social, self-esteem, and self-actualization needs, the higher/growth needs.

Therefore, a "missing link" in Maslow's hierarchy should be added to the pyramid: the need to communicate, inserted between the physiological and security needs. Presented below are the layers of the two pyramids: Maslow's hierarchy of needs and Haberfeld's missing link hierarchy of needs.

Maslow	Haberfeld
Self-actualization	Self-actualization
Esteem	Esteem
Social	Social
Security	Security
Physiological	Communication
	Physiological

The communication needs are broken into two rough subcategories:

1. The *need to talk*, which may include and/or be substituted by other forms of expressions of one's thought process, like drawing, painting, and writing.
2. The more specific form of the need to express the thought process, the *need to vent* your frustrations, to complain, to relieve yourself of unresolved feelings, problems, and dilemmas.

If communication needs are not fulfilled, they will influence other needs, both basic and higher, to the point where dysfunctional behavior will take precedence over any other need, including the need to survive.

Safety or Security Needs

If the physiological needs are relatively well gratified, a new set of needs emerges, which Maslow refers to as safety needs. The human organism may be entirely dominated by them—similar to the physiological needs but in reduced degree. Nevertheless, the safety needs may serve as the almost exclusive organizers of behavior, recruiting all the capacities of the organism in their service, which can then be described as a "safety-seeking" mechanism. Practically everything looks less important than safety. A person in this state, if the situation is extreme enough and chronic enough, may be characterized as living almost for safety alone. To understand clearly the safety needs of an adult, one could look at infants and children. One reason for a clearer threat or danger reaction in infants is that they do not inhibit it at all costs. Thus even when adults do feel their safety to be threatened, it may not be obvious on the surface. The healthy normal, fortunate adult in our culture is largely satisfied in his safety needs. The peaceful, smoothly running, good society ordinarily makes it members feel safe enough from wild animals, extremes of temperature, criminal assault, murder, tyranny, and so on. Therefore, in a very real sense, he no longer has any safety needs as active motivators (Maslow, 1954).

Law enforcement officers, however, are not as free from safety needs as is the rest of our society. Danger is a poorly understood phenomenon of police work. Police officers believe that their work is dangerous, although their perception differs from simplistic media fare. Officers will describe brief moments of terror in the midst of long periods of routine activity. Danger is not sought and not denied but is recognized as an inevitable accompaniment of their work. Danger is a central theme of police work, and thinking about and preparing for danger are central features of the police culture. It is not the actual danger that causes the fear so much as the potential for danger infused in their working environment. Practically anything can happen on the streets (Crank, 1998).

If one adopts Maslow's theory that a person's organism can become wholly dominated by the need for safety, another simple notion should also be recognized. A counseling session, therapy, a peer support group, or any other environment that contains a potential stigma for being weak and fearful will be met with complete resentment on the part of law enforcement officers. They spend their days and nights preparing to deal with the danger to protect others and themselves. A sign of weakness (which will be associated with any reaching out for help—whether external or internal) will immediately decrease the perceived ability to face the danger in a boisterous and forceful way. Officers who are willing to admit that they need the support offered inadvertently admit their weakness. They are stigmatized not so much in the eyes of others but first and foremost, in their own perception. This is the reason that counseling and support sessions in the format offered today are not as effective as they might if approached from a different angle.

Self-Esteem Needs

Maslow describes the overall desire of people for self-esteem:

> All people in our society (with a few pathological exceptions) have a need or desire for a stable, firmly based, usually high evaluation of themselves, for self-respect, or self-esteem, and for the esteem of others. These needs may be classified into two subsidiary sets. First set—the desire for strength, for achievement, for adequacy, for mastery and competence, for confidence in the face of the world, and for independence and freedom. Second set— the desire for reputation and prestige (defining it as respect or esteem from other people), status, dominance, recognition, attention, importance, or appreciation. Thwarting of these needs produces feelings of inferiority, of weakness, and of helplessness (Maslow, 1954, pp. 90–91).

A police officer strives for strength more than does an average person. His or her primary orientation is skewed toward a portrait of strength. One does not picture a law enforcement officer as someone in need of help and/or support— such a picture would defy the entire image of a police officer, an image crucially important to our own safety needs. Needless to say, this image is also crucially important to an officer's own safety needs. The desire for reputation and prestige can be satisfied only if an officer is perceived as strong and invincible. If an officer admits to the need for counseling or any other form of support, this need will turn into a life-long label—*and this label will forever prevent him or her from fulfilling self-esteem needs.*

THE NEW APPROACH

Feelings–Inputs–Tactics Model

It is not our intention here to ignore or reduce the importance of counseling , peer support, or any other stress relief technique that is being offered to law enforcement officers. To the contrary, by introducing the "missing link" in Maslow's theory—the need to communicate and to vent—the crucial significance of the right platform to which to express one's feeling cannot be overstated. However, the key words here are the *right platform*. As stated previously, it appears that we are dealing more with the wrong terminology and approach than with a faulty concept. The terms *support, counseling*, and *stress management* all connote the stigma of being weak, less than able to perform dangerous jobs, perhaps even a danger to others who count on his or her strength during potentially dangerous encounters—in short, less than adequate to be a police officer.

On the other hand, the need to express one's frustrations, fears, dissatisfaction, and an overall sense of injustice is present in police work more than in any other environment. What is therefore the right platform from which to vent feelings, get input from others, and perhaps solicit a tip or two as to how to deal with injustice? Based on years of experience in law enforcement, the only reasonable answer seems to be to build in a mechanism that will not stand out as stigmatic for a given person. In the same way that time is made for officers to participate in biweekly CompStat meetings or roll-call training, time must be set aside for all members of a given agency to participate in meetings during which people take turns revealing their feelings of injustice. Time should be provided for inputs from other participants as well as for tips and tactics on how to deal with injustice in the future. Nobody should be excluded from those meetings or excused for any reason. As for any other mandatory meeting or activity, even members who feel that they have nothing to share with others have to participate, regardless of enthusiasm or willingness. Only by securing the attendance of the entire personnel of a given agency will it be possible to get rid of the stigma and provide a productive and preventive forum. A brief overview of implementation techniques follows.

Implementation Target

Academy Training

It is beyond the scope of this chapter to provide detailed training modules of the FIT model; it should be analyzed and customized by each academy and agency. The significance of the model presented lies not in detailed modules but in introducing a new, quite radical concept that could potentially change the overall morale of police personnel. To destigmatize the idea of stress management, counseling, and peer support, the basic concepts of the FIT model must be introduced during academy training. Officers have to be exposed to the definition of CIS presented above, absorb the potential for encounters in which "justice has not been served," and introduced to the built-in mandatory mechanism of self defense (the FIT) model in the same way that they are introduced to other mechanisms of physical self-defense. Physical self-defense is by no means labeled or stigmatized during academy training. To the contrary, the more able one becomes in physical self-defense techniques, the more one is admired by other officers. There is no reason why the same admiration could not be bestowed upon officers skilled in psychological self-defense techniques.

In-Service Training

Once the FIT model is introduced during academy training, in-service implementation becomes problematic only as far as the actual logistics of the meetings are concerned in relation to available personnel. It is quite obvious that in a smaller agency, the logistics will differ quite significantly from those in a large organization. This is why a detailed module is not feasible;

nevertheless, the following general outline for such a meeting can be customized by each agency.

    I. Meetings are scheduled on a regular basis (in the same way that CompStat meetings or roll-call meetings are). Only in emergency situations will meetings be canceled, and they must be rescheduled within a reasonable time period. The frequency of meetings will depend on the personnel situation in a given agency but should occur not less than once a month.

    2. Emphasis must be placed on the fact that meetings are mandatory for all personnel of a given organization. Nobody will be excluded or excused, no matter how resistant the person is to the idea. In the same way that an officer needs to qualify twice a year or more to maintain firearms or to go through 40 hours of in-house training to maintain his or her certification as a sworn officer, no one will be recertified without attending a certain number of the FIT meetings.

    3. Depending on the organizational culture of the department, meetings can be arranged either by rank or mixed.

    4. No one will to be designated as leader of a given meeting or trained as a counselor or peer support officer. Each meeting will start with somebody who will volunteer to share his or her experience of "injustice" with the others. If no volunteer can be found, officers will draw numbers and the highest number will start. If the first person has nothing to share, the next in line will start (debriefing, stage I).

    5. After the story is shared, feelings about a given "injustice" situation are out in the open. Discussions with others will follow—this constitutes the inputs part. Finally, ideas as to how to deal with a given event, or similar events in the future will be solicited from participants—which constitutes the final part of tactics (debriefing, stage II).

Management Plan for Technical Assistance to Implement the FIT Model

Following is a basic management plan that can be put together to prepare for implementation of this training concept. Although the "ideal" site is identified, based on the size of the agency and the absence of a union (which can sometimes create unnecessary obstacles when new concepts in training are introduced), any size agency (with or without a bargaining unit) can follow most of the steps proposed below.

Although implementation of the plan calls for technical assistance, with minimum resources such assistance can be obtained from local or community colleges, and for the police departments that do not have their own in-house academies, this help can be provided from the regional or in-house training facilities in a given state.

1. Identify the ideal site—a department of between 1500 and 3000 sworn officers (preferably without a union) with an in-house academy.

2. Work with the training director on the introductory modules of the FIT model, to be included during the basic academy training of new recruits. Technical assistance will include:

   (a) Developing an introductory training module

   (b) Identifying and selecting trainers

   (c) Training the trainers

3. Work with the chief, training director, and human relations personnel (and union representatives if a site without a bargaining union cannot be identified). Technical assistance includes:

   (a) Developing a survey instrument to assess the degree of stress exhibited by current employees (survey/questionnaire I)

   (b) Developing a timetable for on-the-job training for every sworn member of the department

   (c) Developing a new operating procedure that would mandate the FIT training model for all sworn members of the department

   (d) Identifying the optimal environment for FIT training

   (e) Developing a survey instrument to measure/assess the degree of stress within the agency after the first year of training (survey/questionnaire II)

   (f) Developing a survey instrument to measure/assess the degree of stress within the agency after the second year of training (survey/questionnaire III)

   (g) Forming focus groups with first-line supervisors to address emerging problems with implementation of and reaction to the FIT model

   (h) Forming focus groups with lieutenants and captains to address emerging concerns with implementation of and reaction to the FIT model

   (i) Establishing a study group to troubleshoot and brainstorm emerging issues/problems with implementation of and reaction to the FIT model (ideas for possible modifications)

   (j) Developing a database program to compile information about stress-related incidents within the department, which will include an early warning system to identify officers in need of more intense FIT training

SUMMARY

The focal point of this chapter was to introduce an alternative approach to stress management training, an approach based on the assumption that the training solutions currently offered are inadequate and misguided. Routinely ignored hidden stressors were introduced and discussed, leading to a new and expanded definition of critical incident stress. Maslow's hierarchy of needs was supplemented with the "missing link" definition, which contributed to a redefined approach to training. The basic concepts embedded in the FIT model do not represent new or innovative ways to stress management. It has been widely recognized that expression of one's thoughts, feelings, and frustrations in front of others is conducive to improved mental health. What is new, and in a way visionary, is a call for implementation of a mandatory platform of exposure for all personnel, one that does not carry a stigma or label. Research projects, in a number of law enforcement agencies, are strongly encouraged to validate the value of this new approach.

REFERENCES

Bouza, A. (1990). *The Police Mystique: An Insider's Look at Cops, Crime, and the Criminal Justice System*. New York: Plenum Press.

Crank, J. P. (1998). *Understanding Police Culture*. Cincinnati, OH: Anderson Publishing.

Finn, P., and Esselman Tomz, J. (1997). *Developing a Law Enforcement Stress Program for Officers and Their Families*. Washington, DC: National Institute of Justice.

Hunt, R. G., and Magenau, J. M. (1993). *Power and the Police Chief: An Institutional and Organizational Analysis*. Thousand Oaks, CA: Sage Publications.

Kureczka, A. W. (1996). Critical incident stress in law enforcement. *FBI Law Enforcement Bulletin*, 65(3), 10–16.

Maslow, A. H. (1954). *Motivation and Personality*. New York: Harper & Brothers.

Maslow, A. H. (1970). *Motivation and Personality*, 2nd ed. New York: Harper & Brothers.

Mitchell, J. T. (1983). When disaster strikes. . . the critical incident stress debriefing process. *Journal of Emergency Medical Services*, 8, 36–39.

Olson, D. T. (1998). Improving deadly force decision making. *FBI Law Enforcement Bulletin*, February, p. 4.

Pierson, T. (1989). Critical incident stress: a serious law enforcement problem. *Police Chief*, February, pp. 32–33.

Sherman, L. (1982). Learning police ethics. *Criminal Justice Ethics, 1*, 10–19.

Sykes, G. W. (1986). Street justice: a moral defense of order maintenance policing. *Justice Quarterly, 3*, 497–512.

Van Maanen, J. (1974). Working the street: a developmental view of police behavior. In H. Jacob (ed.), *The Potential for Reform in Criminal Justice*, Vol. 3. Thousand Oaks, CA: Sage Publications, pp. 83–84, 87, 100–110.

Walker, G. (1990). Crisis care in critical incident debriefing. *Death Studies, 14*, 121–133.

Law Enforcement Leadership Training Community-Oriented Leadership

Introduction

The traditional approach to what can be defined as leadership-related issues in police organizations is replete with terms relating more to management aspects of policing rather than focus on the pure issue of leadership. This chapter, one of the four practical chapters, delineates the problems inherent in underemphasis placed on the need to train police leaders from the very early days of their police careers. Theoretical concepts, together with an overview of some existing training modules, will provide a framework for the development of much-needed training modules to be included at the level of basic academy training.

Leadership Training

A problematic area for many law enforcement agencies is finding potential leaders when a position becomes available. A critical reason that this problem has arisen is because developing and training leadership skills start too late.

Another reason that many departments have a poor pool of candidates is because potential leaders think more about management than about leadership. (The distinction between management and leadership skills is developed further in the Chapter 12). Developing a leader is a long-term process in which encouragement, modeling, and support are combined with natural skills and "the training should begin on the first day of an officer's career" (Bergner, 1998, p. 16).

Many police departments around the country experiment with numerous leadership training modules, some developed and customized by the West Point Military Academy, others by the FBI National Academy, still others endorsed or developed by the International Association of Chiefs of Police. Some departments opt to expose their supervisors to leadership training developed and offered by private consultants or academic institutions, such as colleges and universities with leadership institutes. Others try to customize the available programs and ideas to the needs of the specific environment. Among the leading agencies are the New York, Los Angeles, Washington, DC, Santa Barbara, California, and Boston police departments, and a number of departments in Virginia (Arlington and Alexandria) and other, predominantly large organizations. Despite being clearly dedicated to the development of leadership skills, all the departments continue to ignore the most needy target population—line officers. In their approach to leadership training they continue to utilize the "traditional skills" approach based on the reactive rather than proactive training paradigm. The *reactive training paradigm* focuses on delivering training in various degree of intensity to supervisory employees. The *proactive training paradigm*, advocated by the present author, is leadership training (or any other "new skills" training) delivered very extensively, to line officers during their basic academy training. This proactive training paradigm gains further validity in the light of potential flaws in the civil service system utilized to select police supervisors. While proactive training can serve as a natural screening device and from the onset of their careers identify those who have the greatest potential to assume supervisory positions in the future, this mechanism will not eliminate politically induced promotions. A given agency will still struggle with political interference and pressures from both inside and outside the organization. When the leadership training provided is excellent in both timing and content but the person promoted or appointed is not, there may be potential problems in the future. The political nature of police organizations and the influence it exerts on police leadership cannot be ignored nor underestimated. Nevertheless, providing leadership training to every member of an organization from the very onset of their careers will produce a pool of potential candidates that will be hard to ignore, even by the savviest politicians. As Baker (2001) suggests, the quality of police leadership directly affects the quality of life of police officers and the way they deliver their services; police leaders have the potential to inspire the officers but also to sabotage their efforts. Hopefully, the proactive leadership paradigm will provide more inspiration to police departments around the nation.

LEADERSHIP THEORIES

The numerous leadership theories can be classified, in general, into four categories (Northouse, 2000):

1. "Great man" and genetic theories
2. The traits approach
3. Behavioral explanations
4. Situational theories

The abundance of literature in this field seems to overwhelm the overall approaches utilized by trainers and educators who attempt to integrate various ideas into their leadership training modules. To review, even the most popular of the theories would take away from the criticality of this chapter, which focuses not on a number of leadership theories but on the approach to leadership training in terms of the right timing and degree of readiness.

This author firmly believes that as much as teachers and educators are born and not made, leaders are both. A few leaders are born; more are "made" or developed. Since we have no control over the limited number of born leaders, our obligation to effective community-oriented law enforcement is to "make" more leaders.

According to Swanson et al. (1998), police departments can be divided into three supervisory planes with various levels of leadership skills mix associated with them. The skills mix is based on ranks and consists of three basic skills: human relations, conceptual, and technical.

1. *Top management:* chief, deputy chief, and majors; use predominantly conceptual skills, with marginal use of human relations and technical skills.
2. *Middle management:* captains and lieutenants; use a lot of conceptual skills, but also human relations and technical skills.
3. *First-line supervisors:* sergeants; make relatively little use of conceptual skills, concentrating instead on an equal mix of human relations and technical skills.

The degree to which each of the three tiers utilizes the leadership skills mix might be disputable, based on a number of arguments, such as agency size, political environment, crime problems, economic conditions, and others. However, a much more important issue needs to be addressed. There is a missing link in this equation, and it relates directly to what Bergner (1998) has discussed. The problem of developing police leaders has its roots in full ignorance of the leadership skills that line officers have to utilize on a daily basis. Line officers are the true leaders on the streets. It is they who have to make optimal use of conceptual skills when dealing with the community in any and all daily encounters.

It is of particular importance that police executives and policymakers realize how dependent their agencies are on leadership skills of their line officers (or lack thereof). Human relations and technical skills play a very important role in line officers' work, but in the era of community-oriented policing, the conceptual skills, which are fundamental to problem solving, become the most important of the three skill sets. What was left to middle and top management prior to the advent of community-oriented policing is now placed on the shoulders of line officers. Of course, not all the responsibilities were channeled "down"; however, the idea of decentralization of the chain of command, employee empowerment, problem solving, downsizing and flattening of organizations, and so on, contributed greatly, at least in theory, to the dire need for development of leadership skills of line officers.

If, indeed, leadership skills become so profoundly important for line officers, one would expect to find an extensive leadership training module included in a basic training academy. Unfortunately, this is not the case; most basic training academies not only do not offer an extensive leadership module during basic training but do not include even the most elementary "introduction" to leadership skills.

Swanson et al. (1998) define conceptual skills as the ability to understand and interrelate various parcels of information which are often fuzzy and appear to be unrelated. According to the authors, the standard for handling information becomes less certain and the level of abstraction necessary to handle these parcels becomes greater as one moves upward. The present author does not agree with their interpretation. Today, line officers are supposed to exhibit an entirely new array of skills, they are required to find innovative solutions, involve themselves in conflict resolution, solve problems of a most intricate nature, and so on, face the same if not more complex environments in which the need for well-developed conceptual skills is not a "desired" tool but a necessity. Furthermore, the reactive nature of police work is emphasized much more on the street level than behind a desk. Top and middle management have the luxury of having to make the right or wrong decision in the comfort of their offices, whereas line officers are frequently forced to make a split-second decision in front of a closely observing audience—the community they serve. Such decisions are not simply those in which a police officer's or citizen's life is in real danger, even though such situations are part of the unpredictable reality of life for police officers. Frequently, a police officer has to take control of a situation in which he or she does not have the right leadership skills mix, potentially benign circumstances can escalate to create irreversible damage. The line officer is a true leader in many high-voltage street encounters. It is time to realize this and equip him or her with the necessary tools.

LAW ENFORCEMENT LEADERSHIP

In 1999, during a two-day conference at George Washington's River Farm Estate in Alexandria, Virginia, three dimensions were evaluated as being of primary significance for police leadership in the twenty-first century (Anon., 1999):

1. The changing role of the chief executive
   - Responsibilities to the community
   - Responsibilities to the department/workforce

- Responsibilities to the governing body
- Responsibilities to the profession

2. Preparing for the changing role
   - Community demographics
   - Citizen expectations
   - Community policing
   - Collaboration demands
   - Role redefinition
   - Peer engagement
   - Management style
   - Decision horizons
   - Workforce education levels
   - Perspectives on labor
   - Diversify mandates
   - Litigation trends
   - Evaluation criteria
   - Conflicting demands

3. Sustaining the executive role by managing the external and internal environment successfully
   - Bring passion to the job.
   - Understand yourself and your personal vision.
   - Prioritize customer satisfaction.
   - Become the center of leadership.
   - Lead change throughout the government.
   - Create an environment of partnerships.
   - Continually evaluate change forces.
   - Foster debate and innovation.
   - Approach the thinking process collaboratively.
   - Concentrate on leadership development.
   - Allow others to be recognized.
   - Fashion a diverse organization.
   - Master the vital art of communication.
   - Do not allow yourself to bog down.

Even though the list of new themes certainly applies not only to problems faced by all police officers, not only those in the executive branch, there is no mention of the need to introduce leadership training in the most extensive way during basic academy training.

The ideal list of qualifications and traits for a law enforcement leader were tackled, by many law enforcement professionals, trainers, and educators. Bergner

(1998), himself a law enforcement trainer and consultant, suggested several characteristics that a manager must exhibit to become a successful leader:

1. He or she should show loyalty to the department and his or her superiors, and support any decision they make (if not illegal or unethical).
2. The person is comfortable with a leadership role.
3. He or she is a good example for his or her subordinate.
4. The leader is committed to law enforcement ethics.
5. He or she is a good team player who cares about others.
6. The person has a personal vision that extends to his or her subordinates.

Bergner (1998) further suggested two important areas in developing leadership skills:

1. Creating conditions that encourage and promote good leadership qualities
2. Providing opportunities to act in supervisory and leadership roles

There are several ways in which the first area can be reached successfully. Respect for fellow officers is important, and it motivates a more questionable mindset that promotes better leadership. Another possible condition is creating opportunities for decision making, which can give officers practice in leadership skills. A third condition that may promote good leadership qualities is rewarding teamwork and disciplining lack of teamwork. A fourth condition can be to focus on values and ethics by dealing seriously with ethical dilemmas and bringing in rewards for good ethics (Bergner, 1998).

There are several conditions that can improve the second area. First, the field training officer (FTO) can improve his or her teaching, performance assessing, and feedback skills. Also, the FTO must understand that it is an honor to be an FTO and must understand the responsibility of developing leadership (that it is essential). Second, temporary shift commands are an excellent opportunity for officers to practice leadership skills and for the chief to review an officer's leadership skills. A third scenario that can give officers an opportunity to act as leaders is to have probationers take charge of crime scenes when they are first responders. Fourth, opportunities to train both inside and outside the department are important aspects of good supervision and management (Bergner, 1998).

Griffin (1998) emphasized the importance of evaluating and examining leadership skills. Griffin identified one major problem, pointing to the fact that often leaders have to evaluate their own skills, and suggested a leadership and influence model based on five character traits, the five I's:

1. *Integrity* refers to being able to do the right thing without anybody telling you what to do, and being able to judge right from wrong, good from bad. Leaders with integrity are more competent to judge others fairly and impartially.
2. *Intellect* is often compared to credibility, since it is important that the public understands that the role of a police leader has become increasingly

complex and challenging. It is important to develop a strong intellect for leadership.

3. *Industry* is important for a police leader to possess since it sets an example for how the organization is supposed to be run and at what pace.

4. *Initiative* refers to people who make things happen. Action is an important component of this character trait.

5. *Impact* is when an officer shows confidence, competence, and a positive attitude and makes a difference when he or she arrives at a scene. A leader with positive impact influences others in a good way and thereby makes an organization stronger.

The purpose of the model is to provide a guideline on how to evaluate one's own behavior and how to influence and lead others.

Despite the fact that the ideal characteristics of police leaders were identified by trainers and educators, the major focus in delivering leadership training seems to be misplaced. This problem has two prongs: the audience targeted and the audience's maturity or readiness level. The problem of the audience targeted has already been discussed, in terms of the dire need for extensive leadership training of line officers during basic academy training.

The New York Police Department must be given credit for introducing leadership training to its police officers. However, the three modules that are offered are clearly divided by ranks: Leadership training is offered to sergeants, lieutenants, and captains. The general idea of offering leadership training to three different groups based on rank, is an excellent idea, but the NYPD does not include an extensive leadership training module in its basic academy training, and by ignoring the importance of the timing in introducing its officers to leadership skills, it probably somewhat invalidates the modules offered to its supervisors.

Leadership training is offered to police officers who are already supervisors or, in rare instances, have already been identified as future supervisors. Frequently, however, this has to do with one's seniority on the force or political configuration, and with other skills unrelated to leadership abilities. The goal of this chapter is to offer a reasonable and useful approach to leadership training, not an overview of existing leadership theories. However, the reader is encouraged to use the "Bibliography for the Modules" to become familiar with the relevant concepts.

Although theories and processes that departments employ to identify future leaders are not discussed, it is worthwhile to mention one theory that is of particular relevance to understanding why the existing leadership modules, no matter how extensive and inclusive, still miss the target audience. The effectiveness of any leadership module offered to an audience depends on a certain maturity level, so a theory that emphasizes the importance of the readiness of a given person to become a leader is worthy of special comment.

A necessary supplement to readiness theory (discussed in Chapter 6; the reader is advised to review the readiness levels) is situational leadership theory. Developed by Hersey and Blanchard (1969), it was refined a number of times by the original developers and their followers (Blanchard, et al., 1985, 1993; Hersey and Blanchard, 1977, 1988).

Situational leadership refers to interactions among the following (Hersey and Blanchard, 1993, p. 184):

1. The amount of guidance and direction (task behavior) a leader gives
2. The amount of socioemotional support (relationship behavior) a leader provides
3. The readiness level that followers exhibit in performing a specific task, function, or objective

The situational leadership model consists of four styles. The effectiveness of each style depends on the given situation (Hersey and Blanchard, 1993, p.188):

- *Style 1 (S1):* characterized by above-average amounts of task behavior and below-average amounts of relationship behavior
- *Style 2 (S2):* characterized by above-average amounts of both task and relationship behavior
- *Style 3 (S3):* characterized by above-average amounts of relationship behavior and below-average amounts of task behavior
- *Style 4 (S4):* characterized by below-average amounts of both relationship behavior and task behavior

From these two models, Hersey and Blanchard (1993) proposed a combination theory of what results in the best task behavior and relationship behavior:

- *Readiness level 1(R1).* For learners in R1, S1 leadership is appropriate because these people need high levels of guidance and little supportive behavior. This style has the following descriptors: telling, guiding, directing, and establishing.
- *Readiness level 2 (R2).* The best combination with R2 is S2, because people are trying to do a good job and it is important to be supportive of their commitment. The best descriptors are selling, explaining, clarifying, and persuading.
- *Readiness level 3 (R3).* R3 matches best with S3. The person knows how to perform the task and needs only two-way communication. The best descriptors are participating, encouraging, collaborating, and committing.
- *Readiness level 4 (R4).* For R4, S4 is the best combination. This is because a person at this level is ready to perform, having had enough practice and direction. Useful descriptors are delegating, observing, monitoring, and fulfilling.

Hersey and Blanchard (1993, p. 195) present the following model of best combinations:

- *R1:* S1 high, S2 second, S3 third, S4 low probability
- *R2:* S2 high, S1 second, S3 third, S4 low probability
- *R3:* S3 high, S2 second, S4 third, S1 low probability
- *R4:* S4 high, S3 second, S2 third, S1 low probability

In terms of which people are ready for what jobs, different testing instruments have been developed to measure readiness. Two examples of these are the Manager Rating Scale and the Self-Rating Scale. Both of these measure job readiness (ability) and psychological readiness (willingness).

Hersey and Blanchard (1993) have also developed a model that focuses on the best possible combinations of situational leadership and management. Top management works best with persons in S3 and S4. Middle management can work effectively with any type of leader. Supervisory management is most effective with S1 and S2 (Hersey and Blanchard, 1993, p. 314).

If, indeed, the level of readiness/maturity is clearly related to situational leadership skills—styles of a given person—it has far-reaching implications in identifying the appropriate leadership training for supervisory managers. The training should then be devised based on grouping persons who exhibit certain styles or personality traits and not be offered in an arbitrary manner to police officers holding a certain rank or position in a given agency.

By looking at the three modules devised by the New York City Police Department, where the focus seems to be on rank (e.g., captains, lieutenants, sergeants), it is feasible to assume that these leadership modules might be targeting the wrong audience or that at least some of the potential audience will not derive much benefit from this training concept.

The Pentagon of Police Leadership

The term *pentagon of police leadership* is introduced as a model to be emulated by a basic police training academy. The five components of the pentagon are recruitment, selection, training, supervision, and discipline.

Efforts to develop future police leaders can certainly be traced to the early stages of recruitment and selection. The process of recruitment and selection of recruits with potential to lead, on a street level as well as behind a desk, is just the initial stage of this five-pronged approach to the development of a police leadership cadre. By offering leadership training at the basic academy level, the third prong, it will be possible to identify the level of readiness of officers for future supervisory positions within the department, and further down the road to expose them to advanced leadership training customized to their level of readiness and individual style. The advanced leadership training will then focus on the issues pertinent to the function of supervision and discipline, the fourth and fifth prongs of the pentagon of police leadership.

BASIC LEADERSHIP TRAINING MODULES

Following are two modules of leadership training that the author proposes for adoption by a basic training academy: in-house, regional, or state. The modules include an introduction, goals of the training, instruction processes, detailed outlines, and a bibliography that should probably be updated annually. The modules are referred to as COL (Community-oriented leadership) because the premise behind introducing leadership training at the basic training academy level is to prepare future line officers to deliver the best possible service to community members by equipping them with the necessary leadership skills.

Swope (1998) emphasized the most elementary tenet of leadership, conveying expectations. Neglecting this principle can lead to serious negative effects on both the members of the department and the public they serve. Following Swope's idea, the COL modules are designed to convey the expectations from the very early stage of socialization to police work. Conveying expectations is, indeed, a very critical part of this socialization; however, the *timing* of this process is even more critical. To convey expectations at the time when an officer has already been socialized to a certain police subculture will have little effectiveness on a person's performance.

The first module, the longer one, is appropriate for use at a basic academy that lasts for at least 15 or 16 weeks and can be offered sequentially, each topic once a week for 2 hours. The reason the unit is broken into sequential modules is to introduce the concept at the beginning of the training period and "run" with it throughout the entire session, to enable students not only to absorb the theoretical concepts but to exercise some skills during other instructional modules.

The second model, the shorter version, can be customized for basic police academies that do not offer longer training sessions (e.g., the South Carolina State Academy) but could still fit some fundamental leadership concepts into their basic instruction.

Both modules, but especially the shorter version, can be customized for a different audience, both middle and top management. The amount of time devoted to each topic and the particular focus of each segment can easily be modified to accommodate more experienced students. Furthermore, the readiness/style concepts of the participants should be evaluated prior to modification. The modules were developed based on the leadership training program offered by the West Point Military Academy and were customized to the needs of community-oriented leadership.

Law Enforcement Leadership in the Twenty-First Century: The COL Model

We now explore and develop a conceptual framework of law enforcement leadership in the twenty-first century. The focus is on the complexity of the leadership process in the law enforcement organization from the perspective of the individual members (in their roles as leaders, peers, followers) through the social and work groups to which organizational members belong and in which they work. The complex nature and number of the formal and informal groups that police personnel encounter on a daily basis are discussed and analyzed, together with the environment within which law enforcement organizations attempt to achieve their goals and over which traditional organizational leaders may have little or no control.

Goals of the Course

Historically, all methods, models, and concepts of leadership within a police organization have been based on the implicit assumptions of a homogeneous white male workforce. In the new millennium, most managers and supervisors will need to change their leadership styles to meet the challenges and requirements of a culturally diverse society and workforce. The modern law enforcement leader will have to possess two important traits: vision and the ability to communicate that vision to others, who represent different ethnic, racial, religion, or lifestyle backgrounds, within the workforce and the community (Shusta et al., 1995).

Instructional Procedures

Lectures.   Our specific goals can be met by utilizing the following techniques:

1. During the first meeting, students are divided into working groups. Each group will prepare a short project within the first week of instruction. The project will involve description of the organizational leaders, based on observations from their work environments (e.g., schools, homes, military experience). This project will serve as a working model for the entire semester.

2. To increase student awareness and understanding of the leadership process, various situation scenarios will be provided for analysis in class, based on the concepts discussed in textbooks and articles. Each working group will customize one scenario to meet the leadership styles observed and recorded in their initial project. The final outcome and analysis should demonstrate the desired changes in law enforcement leadership as they pertain to the needs of each specific organization.

3. To increase student understanding of the pluralistic leadership, students are asked to engage in role-playing in which one-half of the group represents the law enforcement leader interacting with various scenarios/situations by the other half. Role reversal is utilized as the follow-up to the first exercise.

4. A model of law enforcement leadership will be developed, COL. The model is based on the existing patterns and is gradually stripped of the unnecessary or obsolete traits, and new and desired features are added as the class progresses into the semester. The final model will integrate the first-week project with the projections for the new millennium.

Videocassettes. Videocassettes containing concrete examples and specific guidance on how to become a more effective leader are presented. A two-cassette audio program, "What Followers Expect from Leaders," contains a unique format of interviews with successful leaders participating in a leadership training session. Thought-provoking questions are posed directly to the listener and can be customized to the specific law enforcement agency from which the student comes.

Guest Lecturers. Law enforcement personnel of diverse background and rank (from line to chief) are invited to deliver short presentations about their perspective on the issues of leadership as they relate to the lessons learned from the past and present and projections for the future.

Course Content

Long-Segment Module

Module I: Law Enforcement Organizational Leadership Model

A. Studying Law Enforcement Leadership as a Science
B. The Scientific Study of Leadership
C. Defining Organizational Leadership
D. Defining the Law Enforcement Organizational Leadership Task
E. Systems Approach to Model Building
F. Model of Organizational Leadership
G. Law Enforcement Organizational Leadership Model versus Reality
H. Summary

Module II: Individual Process and Development

A. Law Enforcement Officer as a System
B. The Dynamic and Diverse Psychological Environment
C. Inputs to the Individual System
D. Throughput Processes and Individual Personality
E. Individual Output: The Production of Behavior
F. The Changing Law Enforcement Officer
G. Summary

Module III: Law Enforcement Leadership: Individual Needs, Expectations, and Motivation

A. Defining Motivation
B. Needs and Motivation
C. Equity and Motivation
D. Expectancy and Motivation
E. Behaviors and Consequences

F. Application of Rewards and Punishment

G. Summary

Module IV: Individual Motivation of a Law Enforcement Leader

A. Organizing a Job through Work Design

B. Two-Factor Approach to Motivation

C. Redesigning the Work Environment

D. Effectiveness and Application of Work Redesign Programs

E. Individual Stress and Adjustment

F. Organizational and Job-Related Stress

G. The Stress and Implication of Life-Threatening Encounters

H. Summary

Module V: Law Enforcement Personnel versus Group Subsystems

A. The Nature of Police Subculture Groups

B. Why Police Officers Join Subculture Groups

C. Structural Dimension of a Police Group

D. What Constitutes Socialization into a Police Group

E. The Goals of Police Socialization

F. Phases of Law Enforcement Socialization

G. Phases of Intervention into the Law Enforcement Socialization Process

H. Summary

Module VI: Processes Within and Between Law Enforcement Groups

A. The Transformation Process

B. Determining Group Effectiveness: An Outside View

C. Inside the Group

D. The Nature of Intergroup Relations

E. Potential for Intergroup Conflict

F. Is Conflict Necessarily Bad?

G. Strategies and Resolutions for Managing Intergroup Relations

H. Summary

Module VII: The Law Enforcement Leadership Subsystem

A. Leadership as a Transaction

B. The Acquisition of Interpersonal Power

C. Leadership Assumptions and the Social Influence Process

D. Personal Attributes and Organizational Outcomes

E. Leader Behavior and Organizational Outcomes

F. Summary

Module VIII: Law Enforcement Leader as a Decision Maker

A. The Nature of Decision Making
B. The Decision-Making Environment
C. Situation Variables and Organizational Outcomes
D. Decision-Making Strategies
E. Impact of Decision Making on the Internal Community
F. Impact of the Decision Making on the External Community
G. Impact of Decision Making on the Law Enforcement Overlapping Identity
H. Assessing Individual Understanding
I. Summary

Module IX: The Law Enforcement Leader as a Communicator

A. Cybernetic Model of Communication
B. The Distortion of Meaning
C. Communication in Law Enforcement Organizations
D. Communication with the External Communities
E. Communication with Overlapping Communities
F. Strategies for Solving Organizational Communication Problems
G. Making the Message Convincing to the Internal Community
H. Making the Message Convincing to the External Community
I. Summary

Module X: The Law Enforcement Leader as a Counselor

A. The Leader–Counselor Role
B. Problem-Centered and Social-Work-Oriented Counseling
C. Performance-Centered Counseling
D. Counseling in the Context of Community Policing
E. Evaluating Your Effectiveness
F. Assessment of the Possible Outcomes
G. Summary

Module XI: Law Enforcement Leader as a Stress Manager

A. Concepts of Stress Management
B. Determining the Source of Organizational Stress
C. Determining the Level of Organizational Stress
D. Stress Management Strategies
E. Managing Stress within the Internal Community
F. Managing Stress within Overlapping Communities
G. Summary

Module XII: Law Enforcement Leadership and Organizational Adaptation

A. Change and Organizational Adaptation
B. Organizational Adaptation to Multicultural Law Enforcement
C. Resistance to Organizational Adaptation: Community Policing
D. Resistance to Organizational Adaptation: Multicultural Communities
E. Proactive Adaptation to Change in Organizations
F. The Open Systems Perspective on Organizations
G. Organizational Throughput Processes
H. Organizational Design Implications for the Leader
I. Summary

Module XIII: Law Enforcement Organizational Leadership and the Ethical Climate

A. Moral Development
B. The Ethical Responsibility of the Leader: Implications for the Organizational Suprasystem
C. The Ethical Responsibility of the Leader: Implications for the External Environment
D. Organizational Influences on Moral Behavior
E. The Consequences of Moral Behavior: Internal Community
F. The Consequences of Moral Behavior: External Community
G. The Consequences of Moral Behavior: Overlapping Communities
H. Summary

Module XIV: Making Sense Out of Perplexity: Law Enforcement Leadership in the Twenty-First Century

A. A Closer Look at the Changing Society
B. A Closer Look at the Changing Demographics of Law Enforcement Organizations
C. Subordinate Expectations
D. Subordinate Performance
E. Making Sense Out of Perplexity: The Leader's Task
F. Projections for the Future
G. Summary

The Short-Segment Module: Leadership for Middle Management and Executives

Week I: Law Enforcement Organizational Leadership Model

A. Studying Law Enforcement Leadership as People-Oriented Science
B. Defining Organizational Leadership as People Oriented

C. Community-Oriented Model of Organizational Leadership

D. Law Enforcement Organizational Leadership Model versus Reality

Week II: People's Perception of the Police: Listening, Responding, and Clarifying

A. Community's Perspective: Us versus Them—An Obsolete Concept?

B. Police Perspective: Us versus Them—An Obsolete Concept?

C. Employee's Perspective: Empowerment—Subordinate versus Supervisor—An Obsolete Concept?

D. Effective Listening versus Effective Responding: Situational Realities

E. Clarifying the Possibilities

Week III: Police Leader as Decision Maker

A. The Nature of Decision Making in Equal-Status Environments

B. The Decision-Making Environment: Control versus Dignity

C. Impact of Decision Making on the Internal and External Communities

D. Impact of Decision Making on the Law Enforcement Overlapping Identity

E. Assessing Individual versus Organizational Understanding

Week IV: Police Organizational Leadership and the Ethical Climate

A. The Ethical Responsibility of the Leader: Implications for the Organizational Suprasystem

B. The Ethical Responsibility of the Leader: Implications on the External Environment

C. Organizational Influences on Moral Behavior

D. The Consequences of Moral Behavior: Internal/External/Overlapping Communities

Week V: Making Sense Out of Perplexity: Police Leadership in the Twenty-First Century

A. A Closer Look at the Changing Society

B. A Closer Look at the Changing Demographics of Law Enforcement Organizations

C. Subordinate Expectations and Performance

D. Making Sense Out of Perplexity: The Leader's Task—Theory

E. Projections for the Future: Assessment of Possible Outcomes

Bibliography for the Modules

Required Texts

Kouzes, J. M., and Posner, B. Z. (1995). *The Leadership Challenge*. San Francisco: Jossey-Bass.
Nahavandi, A. (1997). *The Art and Science of Leadership*. Upper Saddle River, NJ: Prentice Hall.
Northouse, P. G. (2000). *Leadership: Theory and Practice*. Thousand Oaks, CA: Sage Publications.

Additional Required Readings

Avolio, B. J. (1999). *Full Leadership Development*. Thousand Oaks, CA: Sage Publications.

Cogner, J. A. (1989). *The Charismatic Leader*. San Francisco: Jossey-Bass.

Cogner, J. A. (1992). *Learning to Lead*. San Francisco: Jossey-Bass.

Cogner, J. A., Kanungo, R. N., and Associates (1998). *Charismatic Leadership*. San Francisco: Jossey-Bass.

Gilmore, T. N. (1988). *Making a Leadership Change*. San Francisco: Jossey-Bass.

Supporting Bibliography

Bennett, W. W., and Hess, K. M. (1992). *Management and Supervision in Law Enforcement*. New York: West Publishing.

Bennis, W. (1990). *Why Leaders Can't Lead*. San Francisco: Jossey-Bass.

Buchholtz, S., and Roth, T. (1987). *Creating the High-Performance Team*. New York: Wiley.

Chemers, M. M., Oskamp, S., and Costanzo, M. A. (1995). *Diversity in Organizations: New Perspectives for a Changing Workforce*. Thousand Oaks, CA: Sage Publications.

Cleveland, H. (1993). *Birth of a New World: An Open Moment for International Leadership*. San Francisco: Jossey-Bass.

Couper, D. C., and Lobitz, S. H. (1988). Quality leadership: the first step towards quality policing. *Police Chief*, April, pp. 79–84.

Crosby, P. B. (1994). *Completeness: Quality for the 21st Century*. New York: Penguin Books.

Felknes, G. T., and Unsinger, P. C. (1992). *Diversity, Affirmative Action and Law Enforcement*. Springfield, IL: Charles C Thomas.

Hammond, T., and Kleiner, B. (1992). Managing multicultural work environment. *Equal Opportunities International*, 11.

Harris, P. (1990). *High Performance Leadership: Strategies for Maximum Productivity*. Glenview, IL: Scott Foresman.

Jamieson, D., and O'Hara, J. (1991). *Managing Workforce 2000: Gaining the Diversity Advantage*. San Francisco: Jossey-Bass.

Johnson, L. (1991). Job strain among police officers: gender comparisons. *Police Studies*, 14, 12–16.

Kappeler, V. E., Sluder, R. D., and Alpert, G. P. (1995). *Forces of Deviance: Understanding the Dark Side of Policing*. Prospect Heights, IL: Waveland Press.

Kets de Vries Manfred, F. R. (1995). *Life and Death in the Executive Fast Lane*. San Francisco: Jossey-Bass.

Klein, R. (1991). *The Utilization of Police Peer Counselors in Critical Incidents: Critical Incidents in Policing*. Washington, DC: U.S. Government Printing Office.

Klockars, C., and Mastrofski, S. (1990). *Thinking about Police: Contemporary Readings*, 2nd ed. New York: McGraw-Hill.

Kouzes, J. M., and Posner, B. Z. (1995). *Credibility: How Leaders Gain and Lose It, Why People Demand It*. San Francisco: Jossey-Bass.

Mashburn, M. D. (1993). Critical incident counseling. *FBI Law Enforcement Bulletin*, September, pp. 65–68.

Miller, L., and Braswell, M. (1995). *Human Relations and Police Work*, 3rd ed. Prospect Heights, IL: Waveland Press.

More, H. (1992). Male-dominated police culture: reducing the gender gap. In *Special Topics in Policing*. Cincinnati, OH: Anderson Publishing.

O'Toole, J. (1995). *Leading Change: Overcoming the Ideology of Comfort and the Tyranny of Custom*. San Francisco: Josssey-Bass.

Pierson, T. (1989). Critical incident stress: a serious law enforcement problem. *Police Chief*, February, p. 33.

Powell, G. N. (1994). *Gender and Diversity in the Workplace.* Thousand Oaks, CA: Sage Publications.

Roberg, R. R., and Kuykendall, J. (1993). *Police and Society.* Belmont, CA: Wadsworth Publishing.

Shusta, R. M., Levine, D. R., Harris, P. R., and Wong, H. Z. (1995). *Multicultural Law Enforcement: Strategies for Peacekeeping in a Diverse Society.* Upper Saddle River, NJ: Prentice Hall.

Simons G., Vazquez, C., and Harris, P. (1992). *Transcultural Leadership: Empowering the Diverse Workforce.* Houston, TX: Gulf Publishing.

Stojkovic, S., Klofas, J., and Kalinich, D. (1995). *The Administration and Management of Criminal Justice Organizations,* 2nd ed. Prospect Heights, IL: Waveland Press.

Thibault, E. A. (1995). *Proactive Police Management,* 3rd ed. Englewood Cliffs, NJ: Prentice Hall.

Trojanowicz, R., and Bucqueroux, B. (1990). *Community Policing: A Contemporary Perspective.* Cincinnati, OH: Anderson Publishing.

Trojanowicz, R., and Carte, D. (1990). The changing face of America. *FBI Law Enforcement Bulletin, 59,* 6–12.

Walsh, W. (1983). Leadership: a police perspective. *Police Chief,* November, 26–29.

Weaver, G. (1992). Law enforcement in a culturally diverse society. *FBI Law Enforcement Bulletin, 61.*

Relevant Periodical Sources

Academy of Management Journal
Criminology
Equal Opportunities International
FBI Law Enforcement Bulletin
Gender and Society
Gender, Crime and Justice
Journal of Issues
Journal of Police Science and Administration
Justice Quarterly
Police
Police Chief
Police Journal
Police Stress

SUMMARY

The goal of this chapter was to analyze the concepts of leadership, as they represent a step beyond management, and the heart of any unit or law enforcement organization of the future. The more diverse the workforce of law enforcement organization becomes the more leadership is needed. The more dedicated the agency to the philosophy of Community Oriented Policing, the earlier the introduction to leadership skills the better.

Suggestions for new attitudes and approaches based on the assumption that people want to be led and not managed are the desired outcomes of the training modules presented in chapter.

REFERENCES

Anonymous (1999). *Police Chief*, March, pp. 57–60.

Barker, T. E. (2001). *Effective Police Leadership: Beyond Management*. Flushing, NY: Looseleaf Law Publications.

Bergner, L. L. (1998). Developing leaders begins at the beginning. *Police Chief*, November, pp. 16–23.

Blanchard, K., Zigarmi, P., and Zigarmi, D. (1985). *Leadership and the One Minute Manager: Increasing Effectiveness through Situational Leadership*. New York: William Morrow.

Blanchard, K., Zigarmi, D., and Nelson, R. (1993). Situational leadership after 25 years: a retrospective. *Journal of Leadership Studies, 1*(1), 22–36.

Griffin, N. C. (1998). The five I's of police professionalism: a model for front-line leadership. *Police Chief*, November, pp. 24–31.

Hersey, P., and Blanchard, K. H. (1969). Life-cycle theory of leadership. *Training and Development Journal, 23*, 26–34.

Hersey, P., and Blanchard, K. H. (1977). *Management of Organizational Behavior: Utilizing Human Resources*, 3rd ed. Englewood Cliffs, NJ: Prentice Hall.

Hersey, P., and Blanchard, K. H. (1988). *Management of Organizational Behavior: Utilizing Human Resources*, 5th ed. Englewood Cliffs, NJ: Prentice Hall.

Hersey, P., and Blanchard, K. H. (1993). *Management of Organizational Behavior: Utilizing Human Resources*, 6th ed. Upper Saddle River, NJ: Prentice-Hall.

Northouse, P. G. (2000). *Leadership: Theory and Practice*. Thousand Oaks, CA: Sage Publications.

Shusta, R. M., Levine, D. R., Harris, P. R., and Wong, H. Z. (1995). *Multicultural Law Enforcement: Strategies for Peacekeeping in a Diverse Society*. Englewood Cliffs, NJ: Prentice Hall.

Swanson, C. R., Territo, L., and Taylor, R. W. (1998). *Police Administration: Structures, Processes, and Behavior*, 4th ed. Upper Saddle River, NJ: Prentice Hall.

Swope, R. E. (1998). *Conveying Expectations, 65*, 32–34.

COMMUNITY-ORIENTED POLICING
Open Communication Policing

> I don't know the key to success, but the key to failure is trying to please everybody.

INTRODUCTION

This chapter was written to enable police trainers, educators, and academics to design training for community-oriented policing that will explore and analyze the nature of police–community relations as they evolve and change, facing the challenges of maintaining a meaningful relationship with each other. As police–citizen partnership is essential in reducing crime, shaping this partnership in positive ways requires effective police–community relation practice. The major intent is to look at the broader meaning and expectations associated with community policing and the hunger for evaluations of the performance and impact. A short review of the development of this philosophy is offered to serve as a basis for understanding why and how various approaches to the training concepts were developed, how they work, and why the existing formats are not necessarily the ideal models to be replicated in the future. The overview is followed by a summary of some training concepts introduced since the inception of the community-oriented policing philosophy. Finally, based on an analysis of various community policing training modules, a new module of training is proposed to provide an open-communication policing model, one to replace the existing modules, to assist police departments around the country to implement the ideas embedded in the philosophy of community-oriented policing.

ANTECEDENTS OF COMMUNITY-ORIENTED POLICING

Judicial Activism, 1961–1966

The antecedents of community-oriented policing can be traced to what is referred to in the literature as *judicial activism*. Following are some of the major cases of that era.

- *Mapp v. Ohio (1961):* banned the use of evidence seized illegally in criminal cases by applying the Fourth Amendment guarantee against unreasonable search and seizure.
- *Gideon v. Wainwright (1963):* affirmed that equal protection under the Fourteenth Amendment requires that legal counsel be appointed for all indigent defendants in all criminal cases.
- *Escobedo v. Illinois (1964):* affirmed that a suspect is entitled to confer with an attorney as soon as the focus of the police investigation shifts from investigatory to accusatory.
- *Miranda v. Arizona (1966):* required that before questioning suspects, police officers inform them of their constitutional right to remain silent, their right to an attorney, and their right to have an attorney appointed if they cannot afford to hire one. A suspect may waive these rights, but if at any time during questioning the person changes his or her mind as to whether to speak at all or to have an attorney present, the police must respect their decision.

Early Commissions, Foundations, and Programs, 1967–1975

The next period, 1967–1975, is notable for the formation of a number of commissions, foundations, and programs. The commissions were formed in response to urban riots and Vietnam protests. Commission findings opened the door for researchers to analyze police department practices, and led to the formation of the Police Foundation and the Police Executive Research Forum (PERF) (Oliver, 1998).

President's Commission on Law Enforcement and the Administration of Justice, 1967 (The Kerner Commission)

President Lyndon B. Johnson formed the Kerner Commission to investigate the causes of a series of riots in U.S. inner cities during the middle and late 1960s. The Kerner Commission developed five basic suggestions (Williams and Murphy, 1967):

1. Change operations in the inner city to ensure proper officer conduct to eliminate abrasive practices.
2. Provide adequate police protection to inner city residents to eliminate the high level of fear of crime.
3. Create mechanisms through which citizens can obtain effective responses to their grievances.
4. Produce policy guidelines to assist police in avoiding behaviors that would create tension with inner city residents.
5. Develop community support for law enforcement.

Law Enforcement Assistance Administration, 1968

The establishment of the Law Enforcement Assistance Administration (LEAA) grew out of the Kerner Report. The LEAA was a federal agency established under the Department of Justice by the Omnibus Crime Control and Safe Streets Act of 1968 (82 Stat. 197), June 19, 1968. LEAA's functions were to administer grants to government agencies, educational institutions, and private organizations to improve law enforcement. LEAA was abolished, by failure of appropriations, on April 15, 1982. Successor agencies include the Office of Justice Assistance, Research, and Statistics (1982–1984) and the Office of Justice Programs (begins in 1984) (*http://www.nara.gov/guide/rg423.html*).

LEAA was formed to support and educate the police and other members of the criminal justice system in the aftermath of the Kerner Report. One of its goals was to develop a body of knowledge about policing in areas where there was very little (Bratton and Knobler, 1998). During its existence a number of national studies of note were conducted; all focused on the problems of the police and community tension:

- *Report of the National Advisory Commission on Civil Disorders* (Kerner Commission, 1968)
- *To Establish Justice, to Insure Domestic Tranquillity.* (National Advisory Commission on the Causes and Prevention of Violence, 1969)
- *Report of the President's Commission on Campus Unrest* (President's Commission on Campus Unrest, 1970)
- *Police* (National Advisory Commission on Criminal Justice Standards and Goals, 1973)

Police Foundation, 1970

The Police Development Fund, immediately renamed the Police Foundation, was established by the Ford Foundation in 1970. The President of the Ford Foundation, McGeorge Bundy, stated in 1970 that the Police Development Fund was established to encourage innovation and improvement in policing because "the need for reinforcement and change in police work has become more urgent than ever in the last decade because of rising rates of crime, increased resort to violence, and rising tension, in many communities, between disaffected or angry groups and the police" (*http://www.policefoundation.org/docs/history.html*).

New York Community Programs, Early 1970s

The NYPD programs of the early 1970s were not yet community policing but community relations programs: Neighborhood Police Team, Park, Walk & Talk (one officer remains in a radio car while the other officer goes on foot patrol), Cop on the Block (Leibernan, 1972), and other general programs of increased foot patrol.

Team Policing, Early 1970s

Team policing was implemented across the United States in the early 1970s. New York, for example, launched its team policing program in 1971 (Bloch and Specht, 1973). The idea was that officers would be divided into small teams and assigned permanently to small geographic areas. The police would be "generalists," with each officer performing different aspects of police work (patrol, traffic, detective functions) on a rotating basis, although community relations was considered a function of every police officer, regardless of his or her assignment, all of the time (Swanson et al., 1998, p. 14). Team policing was highly regarded at the time of its implementation, but was later deemed a failure (Gay et al., 1977).

Kansas City Preventive Patrol Experiment, 1974

The Kansas City experiment consisted of dividing Kansas City's districts into three groups: One group had strictly reactive beats, where routine patrol was eliminated completely and officers responded only to calls for service; the second group, the control beats, had routine preventive patrol as usual; and the third group had intensified preventive patrol, two to three times its usual level of patrol activity. The study found that the three experimental conditions did not appear to affect crime, crime response time, or citizen satisfaction with police response time or satisfaction with the police. Also questioned was the value of O. W. Wilson's omnipresent policing design, which encouraged the police to use the patrol car to race from to call to call as quickly as possible (Kelling et al., 1974). A later study (Spelmen and Brown, 1981) found that rapid police response is usually unnecessary because most citizens delay calling the police anyway when a crime, especially a serious crime, has occurred.

Police Executive Research Forum, 1975

In 1975, as a response to a rise in crime in the United States and the informative results of some recent research in policing, 10 police executives from large U.S. city police departments met informally to discuss common police concerns.

In 1977, the Police Executive Research Forum (PERF), a national membership organization of police executives from the largest city, county, and state law enforcement agencies, was formed. PERF aims to improve policing and to advance police professionalism through research and involvement in public policy debate. It is funded through government grants and contracts and through partnerships with private foundations and other organizations (*www.policeforum.org/about.html*). Bratton and Knobler (1998) called PERF "an organization of progressive police leaders that conducts research and develops advanced thinking about the police profession" (p. 114).

Later Studies and Programs

RAND Study of Detectives, 1977

Based on the results of a comprehensive survey questionnaire completed by 153 police departments about the workings of their investigation units, the RAND study recommends that police shift their resources from investigative to other resources (e.g., patrol, proactive investigations, public participation programs). The study found that investigators actually spent only 7 percent of their time on activities that actually led to solving crime. The work of patrol officers, accounts of members of the public (victims, witness cooperation), and routine clerical processing led to case solutions in more cases than did the efforts of police investigators. Fifty percent of the investigators' time was found to be spent on postarrest processing, which was inadequate anyway from the prosecution's standpoint. Collection of crime scene evidence was also found to be mostly futile unless the evidence-processing capabilities of police departments were to be improved (Greenwood et al., 1977).

Kelling and Moore (1988) claim that the reform movement came into question in the late 1960s and 1970s. During this period, the impetus for change was spurred primarily by the civil rights movement, migration of minorities into cities, the growing number of youths and teens in the population, increases in crime and public fear, increased oversight of police action by the courts, and the decriminalization and deinstitutionalization movements. In addition to these external social forces, the following also affected the transition to community policing in the United States:

1. The crime rate rose in the 1960s. Police departments had spent a lot of money on improving their resources for new equipment and failed to meet public expectations in controlling or preventing crime. Additionally, policing research in the 1970s found that preventive patrol and rapid response to calls for service did not help to control crime or apprehend criminals.

2. Fear of crime increased. Sometimes the level of fear of crime had no correlation with the amount of actual crime in a neighborhood. Studies in the early 1980s showed that fear had more to do with "disorder" than did crime.

3. Minority residents felt that they were not receiving adequate police service despite attempts by police departments to create equitable police allocation systems and impartial policing to all citizens.

4. The legitimacy of the police was questioned. Student protests and urban riots challenged the police. The U.S. public was viewing the police in action, for the first time on television, in dealing with riots and protests. As a result, the public became more critical of police behavior. Despite police attempts at improved recruitment techniques, women and minorities demanded more representation on the police force.

5. The "myth" of the reform era—little or no police discretion and primacy of "law enforcement" police activities—proved untrue in policing research. Discretion was found to exist at every level of law enforcement and actual law enforcement activities comprised only a small minority of a typical police officer's workday.

6. Patrol officers did not feel respected by police executives during the professional era; in response, militant police unionism rose during this period.

7. The police began to experience a lack of financial and political support from the public and politicians alike. Financial support had been held constant or was increasing in previous years.

8. Competition began to threaten the status of policing. Private security and community crime control movements gained popularity as the public felt they needed more than the police to feel safe.

Newark Foot Patrol Experiment, 1981

A Police Foundation study found that increased foot patrol in an experimental district in Newark did not have a significant effect on crime in the district, but it did lessen district citizens' fear of crime and enhanced their feelings of general personal safety (Anon.,1981). A number of other foot patrol experiments produced similar results; for example, in Newark (Pate et al., 1986) and in Flint, Michigan (Trojanowicz, 1982).

Broken Windows, 1982

The analogy of "broken windows" was used in an article addressing the issue of order maintenance policing, that described citizens' reaction to disorder and crime in their neighborhood. Wilson and Kelling (1982) wrote that "if a window is broken *and is left unrepaired,* all of the rest of the windows will soon be broken ... [O]ne unrepaired broken window is a signal that no one cares, and so breaking more windows costs nothing." Therefore, disorderly behavior signals residents that an area is unsafe and crime is on the rise. If citizens feel that their neighborhood is unsafe, "urban decay" occurs; law-abiding residents withdraw from the neighborhood and weaken the informal positive social controls they presented on the street, leaving the area "vulnerable to criminal invasion," allowing more disorderly behavior to occur in their neighborhoods. The broken windows article relied on the results of the Newark Foot Patrol Experiment (1981) and a number of citizen surveys of public fear to provide support for their broken windows analogy (Wilson and Kelling, 1982). Kelling and Coles (1996) wrote that the broken windows article was influential because it questioned law enforcement agencies' traditional focus on the priority of law enforcement matters. "Recognizing order maintenance by police as significant in reducing fear levels among citizens, and ultimately in preventing crime, lay outside this prevailing concept of police work."

New York Community Patrol Officer Program, 1984

In July 1984, the Community Patrol Officer Program (CPOP) began as a pilot program administered by the Vera Institute of Justice in Brooklyn's 72nd precinct. The aim of the program was to provide a way for the NYPD to provide a community-oriented, problem-solving policing program without major restructuring of its patrol force. The pilot program consisted of 10 patrol officers and one sergeant. There was a 10-day training program for the officers. The aim of the CPOP program was to have every CPO be:

1. *A planner.* Identify priority problems, analyze them, and develop strategies to correct them.
2. *A community organizer.* Work cooperatively with existing organizations on the neighborhood level and stimulate the establishment of such organizations where needed.
3. *An information link.* Provide police with information on, and an understanding of, community problems while informing citizens as to how the NYPD can and cannot be of assistance.

Due to the successes of the program, by 1987 the program had been expanded to all police precincts citywide (McElroy et al., 1990; Weisburd and McElroy, 1988).

Problem-Oriented Policing, Newport News, Virginia, 1987

Under problem-oriented policing, the police should focus on issues that the community sees as a problem. The police attempt to focus on the underlying causes of the problem in the community through a unique four-step process performed by line-level patrol officers: scanning, analysis, response, and assessment (SARA). Police officers are encouraged to create innovative creative solutions to their beat's neighborhood crime problems. This study was sponsored by the National Institute of Justice to see the results of the SARA method in practice in the Newport News police department. The SARA method was shown to decrease crime in Newport News; robberies went down 39 percent, burglaries went down 35 percent, and thefts from parked cars went down 53 percent (Spelman and Eck, 1987).

GROWTH OF COMMUNITY-ORIENTED POLICING

In the early 1980s, the notion of community policing emerged in the United States as the dominant direction for thinking about policing. It was designed to reunite the police with the community. It is a philosophy and not a specific tactic: a proactive, decentralized approach, designed to reduce crime, disorder, and fear of crime, by involving the same officer in the same community on a long-term basis. There is no single program to describe community policing. It has been applied in various forms by police agencies in the United States and abroad and differs according to community needs, politics, and resources available.

Community policing goes much further than being a mere police–community relations program and attempts to address crime control through a working partnership with the community (Peak and Glensor, 1999). This new relationship, based on mutual trust, also suggests that the police serve as a catalyst, challenging people to accept their share of the responsibility for solving their individual problems as well as their share of the responsibility for the overall quality of life in the community (Trojanowicz et. al., 1998).

This new philosophy, however, lacks some serious, in-depth evaluation. Rosenbaum et al. (1994) observe that identification of key factors which affect the creation and implementation of community-policing innovations has eluded systematic study. Literature focusing on community-policing innovations in the United States suggest that two major dimensions of organizational change have been considered by scholars of U.S. policing (e.g., Goldstein 1990; Huber et al., 1993; Kelling and Moore, 1988; Rosenbaum and Lurigio, 1994): external focus and internal focus innovations. External innovations include the reorientation of police operation and crime prevention activities, and the internal innovations involve primarily changes in police management (Zao et al., 1999). The area of adequate training seems to suffer from relative neglect on the part of the researchers preoccupied with field implementations and theoretical themes.

What can be viewed as a utopian perspective of the community and the police, as willing and able or being amenable to change into willing and able to deliver (willing on the part of the community to live the life of true virtue, free of weaknesses and victimless crimes, and able on the part of the police to be able to solve and resolve the crime problems), is illustrated in the 12 points of the COPPS (community-oriented policing and problem solving) philosophy (California Attorney General's Office, 1992):

1. Reassesses who is responsible for public safety and redefines the roles and relationships between the police and the community
2. Requires shared ownership, decision making, and accountability as well as sustained commitment from both the police and the community

3. Establishes new public expectations of and measurement standards for police effectiveness

4. Increases understanding and trust between police and community members

5. Supports community initiative by supplying community members with necessary information and skills, reinforcing their courage and strength, and ensuring them the influence to affect policies and share accountability for outcomes

6. Requires constant flexibility to respond to all emerging issues

7. Requires an ongoing commitment to develop long-term and proactive strategies and programs to address the underlying conditions that cause community problems

8. Requires knowledge of available community resources and how to access and mobilize them and the ability to develop new resources within the community

9. Requires buy-in of the top management of the police and other local government agencies as well as a sustained personal commitment from all levels of management and other key personnel

10. Decentralizes police services/operations/management, relaxes the traditional chain of command, and encourages innovation and creative problem solving

11. Shifts the focus of police work from responding to individual incidents to addressing problems identified by the community as well as by the police

12. *Requires commitment to developing new skills through training*

Conclusions

Not much attention has been devoted by scholars to the problem of effective training that would prepare future peace officers for these new skills. When approaching the issue of effective community policing training, this author was reminded of a play seen a couple of decades ago. The play's title was "The Servant of Two Masters" and can certainly be used as a valid metaphor for the philosophy of community-oriented policing.

As discussed in Chapter 2, traditionally, police forces were utilized to protect those in positions of power, be it emperors, kings, or politicians. Today, in the era of social justice, police are supposed to serve and protect the community. However, despite the decades of wishful thinking and a number of highly publicized research ventures, practically speaking, their deployment today is still predominantly dependent on political considerations. Therefore, if it can be assumed that the idea that has been promoted is to serve the public, there is no notion of proof that at the same time the police force is supposed to abandon its traditional rule. It is true, at least in theory, that a consideration of deployment should reflect the needs and desires of the community they serve. However, many recent and not so recent cases portray a clearly different picture. It is sufficient to mention the case of Elian Gonzales in Miami to illustrate the point that a police agency (or police chief, who in essence represents a given police force) is supposed to cater to the interests of local politicians as much as to comply with federal government decisions—and the two are frequently at odds. If one agrees with this presumption, what is left is to explain is how to become an effective servant to two masters. Open communication policing, the OCP model, may be the vehicle to provide the desired direction.

In addition, the now already old cliché that "there are as many versions of community-oriented policing as there are police chiefs" does not seem to provide an adequate explanation for the lack of effective training, targeting the right audience at the right time. Furthermore, another issue must be recognized and clarified by police administrators and academics. This issue is more rhetorical, or even philosophical, than practical. If a department follows only some of the points (the 12 points of COPPS) or attempts to implement only some of the ideas incorporated in the philosophy of community-oriented policing, is it still committed to this philosophy, or is it just experimenting with some innovative aspects of modern policing?

In the present author's opinion, when a given department proclaims its commitment to the new philosophy of policing but does not follow through with basic tenets or concepts, instead justifying its action using the terms *situational necessities* and *customized approach* (both terms can become very elusive, as one can only imagine), the department is not committed to this philosophy but for various reasons (some having to do with resources, others with personal convictions of the chiefs) claims that it is. In lieu of true commitment, a given department is experimenting with some innovative ideas that may or may not resemble the concepts of community-oriented policing. However, regardless of this author's view, the fact remains that the majority of police departments appear to "swear allegiance" to COP, through various mediums, such as local press conferences, community meetings, various publicized commitments, Web pages, applications for grants, and hiring practices. Therefore, the need to accomplish the public relations agenda, at least partially, remains valid. The only way to fulfill the agenda is to expose recruits to the most extensive and inclusive OCP training at the basic academy level. Prior to the introduction of new modules, a brief overview of the experiments in training of COP is offered to provide for some contrast grounds for the OCP model. The overview is by no means all-inclusive, since as already mentioned, the number of experiments is probably closer to the number of police departments in the United States than to anything else. The overview is limited to a review of existing literature and represents a fairly illustrative mix of current approaches to COP training. It is arranged in a chronological manner to allow an incremental progressive view of experiments.

EXPERIMENTS IN TRAINING

In 1994, the Bureau of Justice Assistance published a monograph with some advice for developing NOP principles in rural police departments. Following are some of the recommendations.

- A planning team is needed to redesign the department's current training needs. Planners may need a course in the planning process (e.g., setting goals, objectives, and strategies).

- In addition to the planning team, local elected officials, officers, city agencies, community-based organizations, and residents would need training in NOP.

- Law enforcement and other service providers would need training in particular problem-solving techniques and community collaboration.

- Possibly there will be a need for specific training under NOP (e.g., training for school teachers to detect drug use in teenagers).

- Department planners will need to assess whether it is cost-effective to have in-house trainers for some or all NOP training sessions.

- It might be more appropriate to send certain personnel for more specialized training at regional institutes or in another jurisdiction's academy.

Savannah, Georgia, Training Program

In 1995, McLaughlin and Donahue followed the curriculum of the Savannah, Georgia, Police Department (SPD)'s transition to community policing. The recruits were trained in seven training modules, five of which were completed when their article was written. Since then the department has completed all the modules. Savannah has approximately 500 employees (Oliver, 2000).

Module I: Participatory Decision-Making and Leadership Techniques for Management, Supervision, and Street Officers.
Module I is designed for training police administrators and managers exclusively. An overview of COP is taught, as well as the COP features that should be taught to employees. A bibliography of books and articles is given to participants to read on their own. Politicians are invited to attend the training seminars as well.

Module II: Community-Oriented Policing.
Module II offers an overview of COP to SPD employees. The students are given a copy of the executive summary of the new precinct system that is an important feature of the COP program (the new precinct system is based on data from criminal justice and social services agencies on the type of police organization that would decrease crime in a community). Copies of "Broken Windows" (Wilson and Kelling, 1982) and *Crime and Policing* (Moore, et al., 1988) were also handed out to the participants. The role of managers as COP resources and decision-making authority figures is highlighted. The role of officers as problem solvers is encouraged. There is a focus on the steps that the SPD has taken toward adoption of the new COP philosophy. The instructional staff emphasizes that successful integration of COP into the department will take several years to occur within the SPD.

Module III: Problem-Oriented Policing.   In module III the objectives and outlines of POP procedure are explained. *Problem Oriented Policing* (Spelman and Eck, 1987) is assigned reading. The SARA method is explained to the officers in full detail. The officers deal with POP problem-solving exercises and watch a video on the subject by the Fort Meyers (FL) police department.

Module IV: Referral System, Materials, and City Ordinances.   Module IV encourages police officers' use of referrals to social service agencies and the use of specific city ordinances to deal with problems.

Module V: Developing Sources of Human Information.   Module V stresses the importance of contact with all people on a continuous basis, including the use of informants, field interviews, and investigator detentions.

Module VI: Neighborhood Meetings, Survey of Citizen Needs, and Tactical Crime Analysis.   Module VI reviews methods of soliciting citizen input. In addition, it reviews methods of crime analysis.

Module VII: Crime Prevention and Home Surveys.   The aim of module VII is to review with officers how to conduct security surveys of home and businesses and summarizes and integrates concepts from all seven modules (Oliver, 2000).

Survey of Chiefs of Police

In 1995, Zhao et al. published an article that was based on data derived from a national survey of police chiefs conducted by Washington State University. The surveys were conducted in three-year intervals since 1978 with police chiefs who served a population of over 25,000 people. The sample includes 281 municipal police departments in 47 states. In 1993, there were 228 respondents to the survey. The survey found that police departments across the nation had similar interests in COP training and education (training in ethics, police–community relations, principles of COP). It was found that improving police community relations was top priority for the police chiefs surveyed in 1993.

New Haven, Connecticut, Training Program

In 1996, Codish described the New Haven, Connecticut, Police Department's experience with community policing training implementation.

1990–1992.   New Chief Nicholas Pastore decided to revamp the current police academy to focus on community policing. The chief began a search for a civilian education director with academic credentials with ties to community groups. He empowered the director of the academy to report only to the chief. Pastore returned the eight sworn instructors in the current paramilitary academy to patrol duty.

1992. Codish himself was appointed as the new director of training. He decided that the academy would have the structure of a university. A new student handbook was issued that stressed academics, original research, communication skills, critical thinking, community involvement, and officer discretion and dialogue. Disciplinary calisthenics was replaced with assignments in research and writing. The new curriculum attempted to get the *students* (no longer *recruits*) into the community as much as possible (e.g., volunteering at a soup kitchen) and changed militaristic or sexist terms (e.g., *police department* instead of *force, officers* instead of *men, staffing* instead of *manning*). Seminars with outside consultants were added to the curriculum, including professors from Yale University's law and medical schools, the Anti-Defamation League, and the AIDS project of New Haven. Peer and faculty evaluation components were included. Physical training was eliminated. The passing academy score of 70 percent was raised, that of the Police Officer Standards and Training Council, to 80 percent for graduation from the academy.

1993. Some conclusions regarding the nature of the training were reached.

1. The length of the training, 17 consecutive weeks of academy training was too long; however, 40 consecutive hours was too little.
2. Teaching must be made as experiential, interactive and participatory as possible.
3. Students must be able to read, speak, and write well.
4. Students must be committed to community policing and working with the community.
5. Racial, cultural, and ideological diversity among the students and faculty is encouraged.
6. Physical training is extremely important for promoting fitness, discipline, and teamwork.
7. Attitude and demeanor are key, demanding more in-depth psychological screening.
8. The relationship between creative problem solving and the ability to follow rote orders needs further study.

1993–1994. Major academy revisions were put on hold while academy programs were developed, including in-service programs and university student internships in the police department.

1994. The academy took on the new role of outreach to attract recruits. The department put ads in church and synagogue bulletins, flyers at social and religious events for minorities, and in advertisements in English- and Spanish-speaking media.

1995–1996. More interactive teaching techniques were incorporated into the academy curriculum (e.g., a discussion of search and seizure would be interrupted by two staff members casually entering and having a conversation with the instructor while surreptitiously removing a watch from the desk before leaving).

1996 and Beyond.   In five years the academy hoped to become a two- and four-year degree-granting college. The broad goals of the department were identified:

1. To raise the educational level of police personnel
2. To increase the number of women and minorities in the department
3. To take advantage of community resources and make police resources available to the community
4. To encourage and facilitate continuing education through Yale's School of Management, Child Study Center, School of Law, School of Public Health, Department of Psychiatry, and the Institute for Social and Policy Studies

REGIONAL COMMUNITY POLICING INSTITUTES

The same year, 1996, marked a very important period in the development of community policing training. A request for proposal was issued by the Office of Community-Oriented Police Services for applicants to design regional community policing institutes (RPIs). The proposal emphasized the need to design training modules based on partnerships between police departments and academic institutions (Anon., 1996).

Regional partnerships needed to provide comprehensive and innovative education, training, and technical assistance to COPS grantees and other policing agencies throughout a region. A region is generally considered a state or multi-state collaboration. The RPIs would have the latitude to experiment with new ideas to challenge and improve COP programs. There is also the goal of helping agencies sustain COP once government funding has ceased. The agencies qualified for COP were identified as state/regional training academies, local county police training facilities, police officer standards training commissions, and universities/colleges. There is the need to show COP programs in the last two yeas. Awards of up to $1 million were made to the grantees.

The request facilitated the establishment of 35 institutes nationwide that were originally funded for one year, with the possibility of renewal for two additional years (Oliver, 2000). Four years later most of the originally founded institutes still exist and offer a similar module, which is usually delivered during two weeks and is divided into 10 or 11 training segments.

Following is a sample of a curriculum offered by one of those RPIs, the New Jersey Regional Community Policing Institute.

- *Module 101:* Orientation to Community Policing
    - Definitions and history of COP
    - Philosophy and concepts
- *Module 102:* Cultural Diversity in Community Policing
    - Tools to learn, practice, and implement an appreciation of diversity

- *Module 103:* Community Organizing and Partnerships
  - Key steps in building effective community policing partnerships
- *Module 104:* Conflict Resolution in Community Policing
  - Basic understanding of crisis and conflict resolution
- *Module 105:* Crime Prevention in Community Policing
  - Basic concepts of crime prevention as they relate to the practice of community policing
  - Information about the Capstone Project
- *Module 106:* Community Resources and Services
  - Police officer's roles of referral and networking with public and nonprofit agencies
  - Introduction and familiarization with the range of services and agencies
- *Module 107:* Problem-Solving Strategies in Community Policing
  - Components of problem solving
  - The SARA model is presented through lectures, case studies, and interactive exercises
- *Module 108:* Strategic Planning in Community Policing
  - The value and role of strategic planning
- *Module 109:* Community Policing Program Management/Research Skills
  - Methods of design, implementation, and evaluation of community policing
  - The Capstone Project is discussed
- *Module 110:* Ethical Issues in Community Policing
  - Historical perspective of police integrity
  - Ethics in the context of the new roles and responsibilities

Tables with additional curricula topics and hours allocation, from numerous RCPIs around the country can be found in Table 9–1. The variety, in both content and length of the modules, illustrate, one more time, the critical issues embedded in the currently offered Community Oriented Policing training, and a dire need for a generic, standardized outline.

Despite the fact that the idea of creating these regional institutes is definitely worth a commendation since it provides for much-needed uniformity in training certain basic concepts of police work, a number of problems need to be addressed generically. It is possible that some institutes address the problems discussed in this chapter; however, the majority seem to fall into the same trap.

The topics included in the curricula are all important and significant, but no matter how important a topic or a theme, it needs a proper introduction, and a correct sequence, without which the importance of the entire module seems to lose the desired impact.

Table 9–1 Regional Community Policing Institutes

Course	Hours	Type
Arizona		
Community policing		Separate
Customer service	8	
Personal accountability	5	
Problem solving for police	8	
Creating partnerships in the community	4	
Ethics	6	
Cultural awareness	8	
Interpersonal relations skills		
Conflict resolution	8	
Fundamentals of crime prevention	8	
Creating community-oriented field training	4	
Futuristics	4	
Los Angeles, California		
Introduction to community policing	8–24	County sheriff
Community policing for supervisors and managers	16	
Criminal abatement workshop	24	
Practical skills for crime prevention	24	
Community responses to youth crime	16	
Sacramento, California		
Basic community-oriented policing and problem solving	24	Police
Train the trainer	40	
Effective supervision of problem solvers	16	
Neighborhood revitalization and advanced problem-solving course	8	
San Diego, California		
Community policing and problem solving	3, 4	Police
Mediation and conflict resolution	4	
Crime prevention through environmental design	4	
Crime-free multihousing program	4	
Homeless outreach	2	
Grant writing	2.5	
Basic facilitation skills	3	
Code compliance	3	
Nuisance abatement	3	
Mediation skills	8	
Crime mapping	4	
Problem solving for community-oriented government	4	

Table 9–1 Regional Community Policing Institutes *(continued)*

Course	Hours	Type
Colorado		
Basic community policing	16	State Division
Facilitation skills—community policing partnerships	16	of Public safety
Executive leadership series	32	
Strategic planning	8	
Advanced community policing series	8	
St. Petersburg, Florida		
Introduction to community policing	16	College
Police–community partnerships	16	
Problem solving for the community police officer and citizen	16	
Survival skills for community policing officers	24	
Ethical issues and decisions in law enforcement	8	
Reaching your goals through goal compliance	8	
Crime prevention/crime displacement and environmental design	16	
Managing encounters with the mentally ill	16	
Community policing for citizens	4	
Managerial buy-in	16	
Strategic selection, management, team building, and retention of community policing	16	
Managing organizational change	16	
Protecting, serving, and supervising through community partnerships	8	
Fort Wayne, Indiana		
Implementing community policing	8	Separate
Implementing community policing—technical	4	
Community policing and change theory	4, 8	
Establishing collaborative community partnerships	8	
Inventory conference (community policing)	16	
Conducting community surveys—technical	2, 4	
SARA problem solving	8	
Train the trainer	40	
Strategic planning for LE agencies	8	
Ethics—technical	2, 4	
Ethics	8	
Diversity for law enforcement officers	8	
Survival Spanish	24	
Graffiti and your community	8	
COP and gangs	8	
COP and female gangs	8	

Table 9–1   Regional Community Policing Institutes *(continued)*

Course	Hours	Type
Kansas		
Ethics and integrity		College
Community policing implementation		
Contemporary issues (school violence, illicit drugs)		
Train the trainer		
Kentucky		
Making communities safer	32	College
Managing change	16	
Domestic violence/adult sexual assault	40	
Criminal justice executive development course	320	
Louisiana		
Introduction to COP	8	Separate
COP problem solving	8	
COP for diverse populations	8	
Effective communication skills	8	
Effective conflict resolution	8	
Peace officers' ethics and integrity	8	
Survival skills for peace officers	8	
Neighborhood leadership and development	4–8	
Community mobilization and partnership	4–8	
School violence prevention	8–16	
Grant writing and resources for community policing	8	
Domestic violence; peace officer response	16	
101 effective principles of leadership and supervision	4–8	
Leadership and supervision	4–8	
Leaders training		
Effective management of COPS training		
Michigan		
Community policing	16	College
Community policing initiatives	12	
Supervision and community policing	16	
Crime and data analysis	8	
Problem-solving workshop	8	

Table 9–1   Regional Community Policing Institutes *(continued)*

Course	Hours	Type
North Carolina		
Introduction to community policing	8	Separate
Changing the role of supervisors	16	
Applied police skills	16	
Integrity and diversity (CP)	16	
Leadership skills—CP		
Leadership skills—Managers/supervisors	24	
Accountability training—Supervisors	16	
Problem solving for property managers	8	
Problem-solving template	8	
Ethics and integrity—police managers	16	
Crime mapping—CP	16	
Introduction to GIA—analysts	8	
Intermediate GIS—analysts	16	
Advanced GIS—analysts	8	
New Jersey		
Orientation to CP	7	Separate
Cultural diversity—CP	7	
Community organizing and partnerships	7	
Conflict resolution—CP	7	
Crime prevention in community policing	7	
Community resources and services	7	
Problem solving and strategies in CP	7	
Strategic planning—CP	7	
CP—management/research skills	7	
Ethical issues	7	
Middle management training conference—CP	16	
Chief's executive training program—CO	16	
Advanced level training program (instructors)	16	
New York		
		College
Oklahoma		
Introduction to CP	8	Separate
Problem solving	8	
Human relations for LE	8	
Marketing for LE	8	
Leadership skills for CP coalitions	8	
Ethics in CP	8	

Table 9–1   Regional Community Policing Institutes *(continued)*

Course	Hours	Type
Tennessee		
Executive briefings—COP	30–40	
Executive briefings—community connection to CP	30–40	
Managing organizatonal change	30–40	
Implementing agency-wide COP	30–40	
Practical approaches to problem solving	30–40	
The supervisor and COP	30–40	
Managing patrol time and crime analysis	30–40	
Texas (University of Texas–Austin)		
Community policing and problem solving	24	College
What is in it for you?	4	
Conflict prevention and management	16	
Ethical leadership	24	
Texas (Sam Houston University)		
Keeping the community in COP	8	College
Technological issues	16	
Community policing and problem solving	8	
Implementing community policing	16	
Ethics and integrity	8	
Supervising in a CP environment	8	
Washington		
Introduction to CP and problem solving	2–16	Separate
Community partnerships and effective meetings	4–16	
Curricula development	4–16	
Problem solving and facilitation skills	6–16	
GIS	16	
Cultural awareness and communications	1–8	
Officer survival	2–4	

The OCP module proposed by the author addresses most of the deficiencies identified in the RPI modules, and the reader is directed to the last section of this chapter for further discussion of the desired themes and sequences.

Training Studies

In 1997, three studies, each analyzing a different training environment—large municipality, small rural city, and one state's training initiative—were published. Each study had its contribution to a better understanding of the flaws and desires in COP training.

Aragon and Adams (1997) analyzed the experience of the Whiteville, North Carolina, Police Department within the context of the department's community policing adoption. Whiteville is a small southern city located in Columbus County in the southeastern section of North Carolina. There are 5600 residents in Whiteville, 35 percent of whom are members of minority groups.

Whiteville COP training has two levels: basic and advanced. The basic level consists of (1) a classroom presentation, (2) 4–6 hours of on-the-job training with a COP officer, (3) 1 hour of bicycle training, (4) a written test, and (5) interviews with COP team leaders. By the end of their training, officers learn how to make introductory visits, work with community members, and solve problems using the SARA model. The advanced level consists of two courses (2 days each) at the North Carolina Justice Academy. The courses at the academy focus on leadership styles, total quality management, and COP support. Officers receive tabs for their uniforms upon completion of each course.

The authors recommended that departments provide at least rudimentary in-service COP training until officers can attend formal police academy courses. Officers should receive awards for their participation in the COP philosophy.

Breci (1997) surveyed a random sample of 1526 Minnesota peace officers. Fifty-one percent ($n = 801$) of the surveys were returned. The assumption was that organizations can help the transition to COP in two ways: (1) the development of continuing education courses that focus on the development of officer competence in those areas that are central to COP success, and (2) the financial support of departments toward officers' college tuition.

The findings of the survey suggested that police departments in Minnesota are not adequately encouraging their transition to COP under two measures of analysis: agencies that provided financial support for education ($n = 46$) compared to ($n = 123$) of those that did not. A low number ($n = 43$) of agencies offered tuition reimbursement compared with a high number ($n = 120$) that did not. In addition, the percentage of officers who attended various training courses (in the three years prior to the study) were exposed to a relatively low percentage of courses that offered COP-related skills (firearms, 92 percent; defensive tactics, 76 percent; human relations/human behavior, 42 percent; communication skills, 37 percent).

Skogan and Hartnett (1997) described the experience of the Chicago Police Department in implementing their new training modules. Chicago's community policing programs are known as CAPS. The CAPS training consisted of an initial orientation and then follow-up skill-building sessions. The skill-building session lasted three days. The sessions consisted of an overview of COP principles. At the end of each session, there were question-and-answer (Q&A) sessions between officers and senior police officers. The Q&A was aimed at COP issues but turned primarily on other non-COP issues. The skill-building sessions lasted three days. They were taught by a mixture of sworn and civilian trainers. The sessions took place outside the police academy, police wore civilian clothing, and the officers were taught in mixed-rank groups. It was found that there was not enough time for preparation for CAP training before the training had to begin. It was felt that there was not sufficient time for trainers to revise their curriculum in accordance with COP sources outside the department.

A lack of unity among police personnel posed a problem for the successful integration of CAPS. Many officers expressed hostility toward the CAPS curriculum, which they felt was created by a civilian consulting firm. They felt that the civilian trainers did not understand police operations. In fact, sworn trainers also resented civilian trainers. The sworn trainers felt that they had put a lot of work into designing the curriculum. Their suggestions were mostly ignored by the consultants. Police supervisors criticized the mixed-rank groupings. They did not like to be debated by their subordinates. They feared that the casual atmosphere might lead to their subordinates' dismissal of their commands on the street.

The CAPS curriculum was supposed to foster active learning, but the recruits ended up being lectured to in the traditional classroom manner anyway. Many officers interviewed by the authors related that the CAPS lectures were too abstract and that they could not realistically put the CAPS principles into everyday use. Officers also believed the contention that COP has been a failure in other cities. In the months that followed CAPS training, the training was revamped with two experienced consultants, including the acting director of a big city police department. The redesign team concluded that the initial curriculum was too abstract and lacked clear guidance for officers. The training had to begin with COP material that was clear and straightforward and move on later to the more abstract. The decision was made not to mix ranks in the classroom. The redesign team suggested that trainers announced that there would be a test at the end of the CAPS training. If the officers did not pass the test, they would have to repeat the training. The test component increased officer attentiveness. (Some other deficiencies of CAPS training were discussed in earlier chapters.)

Finally, some recent very interesting developments can be observed. From the totally oblivious to the most progressive, from Anchorage, Alaska, to Washington, DC, the picture looks either very dark or very bright, depending on what side of the road one stands on. Riley (1999) described the COP implementation in Anchorage, Alaska. Anchorage has a population of 255,000, but the COP effort was implemented only in one experimental area, Mountain View. Mountain View has a population of 5600 people. This area was one of the city's more high-crime districts.

The author based his recommendations on the data of interviews of 23 officers on the subject of COP training. The study concluded that departments needed clearer articulation of COP goals and greater attention to training and orientation for new members and clearer standards for evaluation of COP team members. Only 5 of the 23 officers interviewed could articulate the definition of community policing. That was clear and consistent with the one taught at their COP training. Despite an official statement of COP philosophy posted at the Mountain substation, 18 of the 23 officers were unable to quote or roughly paraphrase the statement.

In 1999, the *FBI Bulletin* published an article designed to provide some general advice to police departments that want to integrate more COP philosophy into their curriculums (Anon., 1999).

1. Review the current curriculum.

2. Make sure that the curriculum does not teach mostly what recruits end up spending a small amount of their time doing in practice (e.g., law, department regulations, skills).

3. The academy should focus on interpersonal and communication skills and problem-solving skills needed in a COP atmosphere. These courses should include courses in ethnic diversity, drug alcohol awareness, and verbal de-escalation skills (e.g., verbal judo).

4. After the instructional methods. The lecture method of training should be altered and the emphasis on mastery and obedience should be shifted to a focus on officer empowerment. Foster self-directed learning. Create small work groups in the classroom where officers are given the tasks of solving problems that might appear in the community. In this aim it is helpful for instructors to teach in the *andragogy* (teaching of adults) style instead of *pedagogy* teaching techniques. An example of andragogy would be having recruits discuss in small groups their personal experiences of race and diversity.

5. Encourage ethnic diversity among the academy faculty and student body.

Harvey (1999) recognized the role of the FTO in building support for community policing, in an article published in *Police* magazine.

The article contains advice concerning the adoption of COP into police recruit training. First, there should be classroom training on COP, consisting of an explanation of current styles and trends of COP, departmental history of COP, and *how and why* it was adopted. Failed attempts at COP should not be omitted. Explanations should be offered as to whether the department employs a whole departmental model or a specialized unit model of COP integration. Included should also be some presentations of area COP *gurus*.

In the FTO component of training, firsthand accounts that display the real-life impact of COP should be provided. Recruits should be exposed to various neighborhoods where COP is in place and attend community meetings. Recruits should also be introduced to citizens who can personally tell officers what it was like before the department's COP implementation. Finally, recruits should participate in a small POP project. Once in the field, it is the first-line supervisor (sergeant) who will have the most impact on the new officers. The sergeant has the responsibility for keeping up a positive COP atmosphere for new officers. They should closely monitor the curriculum program and adjust for the necessary changes.

Finally, an innovative approach was adopted by the Metropolitan Police Department of Washington, DC. As a part of their academy training, recruits visit and participate in a workshop at the U.S. Holocaust Memorial Museum in Washington, DC. The idea was initiated by Police Chief Charles Ramsey and the Anti-Defamation League. The recruits have a tour of the museum and participate in educational sessions. The recruits are given case studies that look at the choices the police made simply following department orders. The aim of the workshop is to have recruits consider the protective rights of our Constitution (Anon., 2000).

(This idea is especially noteworthy given the fact that their basic academy training is still skewed very heavily toward the traditional skills, as discussed in Chapter 4).

Critique of Training Programs

Despite the numerous ways and attempts to deliver adequate COP training, the picture looks similar to the approach advocating the proactive training paradigm for leadership training. The major problems with existing attempts to deliver effective community-oriented training are the timing, content, and intensity of the message conveyed. During their basic academy training recruits must be exposed to intensive and extensive community-oriented training in the OCP model, based on the open communication component. This component refers to the basic premise upon which any COP training has to be based. This concept was developed by the present author based on the assumption that the message conveyed to both the public and police officers has to be one based on an understanding that police agencies are at best servants of two masters. This concept should be developed much further during the academy training to emphasize the elusiveness of the word *community* or *public*. The concept should not only be explained starting with an overview of the police role in the society, but articulated to the point that officers will be comfortable discussing their ambiguous role with members of the community. Partnership and cooperation must be based first and foremost on full understanding of what both parts are able to deliver. The impediment of being a servant to many masters has to be made clear to both the public and the police. The role of educating the public can be placed only partially on the shoulders of the police; the rest of the burden is the responsibility of our school systems and the media. A police officer who is not aware of his or her ambiguous role cannot proceed to understand the topics introduced by the RPI modules.

The pressures exercised by the so-called "community" on any police agency amount to numerous unattainable demands. The elusiveness of other terms, such as *problem solving, joint ownership, joint responsibility,* and *cooperation* (as listed in the 12 tenets of COP) has to be deciphered from the very early stages of socialization to police work.

Less emphasis should be placed on problem-solving and conflict resolution skills. If we could teach effectively the skills needed to resolve conflicts and mediate between parties, we would not need any more meetings at Camp David to resolve the Middle Eastern conflict or the one in Northern Ireland. Instead, the focus of training should be placed on our fears and weaknesses: understanding of the dynamics between police officers and parents of delinquent children, medical conditions of mentally ill people, problems of adolescence, jealousy, hate, and ethnocentrism; and our insatiable demand for drugs, gambling, prostitution, and our propensity for other "victimless" crimes. The problems listed above have their roots in the history of humankind and are not going to disappear from our life with the skillful intervention of police officers working together with the community.

A number of police departments around the country pride themselves in their ability to "solve the problem" of prostitution, for example. The author has serious doubts about how this problem was really solved: Are we to believe that the women involved in this oldest profession were actually convinced to change their profession? The issue of crime displacement has been researched by many scholars and the results of those studies are not very encouraging.

OPEN COMMUNICATION POLICING MODULE

It is true that police officers can make a difference, especially when cooperating with the public. The key to understanding this difference, however, and the scope of the difference can be found in the open communication policing (OCP) module—when police officers are trained to be open about what they can and cannot solve, regardless of the amount of cooperation from the public. To advocate in the utopian 12 points of COPPS for "shared ownership" and "accountability" as well as "sustained commitment" from both the police and the public places the most unreasonable burden on the police force.

The need for in-service training through regional training institutes will remain valid until all the "traditionally" trained officers retire. Even then, the existence of these institutes will be of vital importance in delivering updates and developmental training regarding emerging issues in police–community relations. For now, though, police trainers, educators, and policymakers must concentrate on delivering adequate funds not only to the regional institutes but first and foremost to in-house, regional, and state basic police academies.

One of the main reasons that this author advocates extensive basic academy training in OPC has its roots in the *equal-status contact hypothesis*. The new intersections of class, race, gender, and the social justice issue and the concepts they bring to the U.S. society can be comprehended only in a closed-contact environment. Even then, this comprehension will be extremely limited, but looking at the full half of the glass (rather than the empty half), one can clearly see the potential inherent in this environment versus the existing options.

Allport (1954) identified 25 variables, organized into six general areas, which can affect the reactions, experiences, and prejudices of the participants in intergroup contacts. Further research in the area identified four factors that are especially influential in the interactions between groups:

1. Equal status
2. Intensive interaction
3. Noncompetitive relations
4. Cooperative tasks

The equal-status contact hypothesis endorses the idea of the presence of all four of these factors in an intergroup situation as a catalyst to better understanding of each member of the group. The critical importance of the equal-status contact hypothesis will be revisited in Chapter 10. For now it is sufficient to mention that emphatic qualities and understanding of the community can be developed in the most effective way by socializing recruits, from the very early stages of their training, to the idea of cooperation and problem solving of the unknown. The term *unknown* is used loosely to allude to the problems that officers will face on the streets on a daily basis while policing the tremendously diverse and opinionated public of the new millennium.

Training Program

Required Text.   D. H. Ronald, P. D. Mayhall, and T. Barker, *Police–Community Relations and the Administration of Justice*, 5th ed. (Upper Saddle River, NJ: Prentice Hall, 2000).

Recommended Text.   K. J. Peak, and R. W. Glensor, *Community Policing and Problem Solving: Strategies and Practices*. (Upper Saddle River, NJ: Prentice Hall, 1999).

The training starts by identifying the background of recruits and dividing them into working groups that are as diverse as possible. The Capstone Project is introduced prior to the first training module. The Capstone Project is structured in such a way that only full cooperation between the members of a given group with the members of all the other groups will allow them to achieve the desired outcome or, in other words, to solve the problem. The true validity of the test will be measured not necessarily by its successful completion but by the efforts and innovative techniques and solutions employed in the process.

- *Module I:* Introduction: Why Open Communication Policing? (Broken promises are more destructive than broken windows)
- *Module II:* Police—Community Relations: The History of Policing (One master's servants)
- *Module III:* The Public and the Police: Theory and Practice (The public are not the police and the police are not the public)
- *Module IV:* Police Role in a Changing Society (Two masters' servants)
- *Module V:* The Communication Process (How close is too close?)
- *Module VI:* Police Discretion and Community Relations (Will you still respect me the next morning?)
- *Module VII:* Community Policing and Problem-Oriented Policing (Drugs, gambling, prostitution: Big Brother knows better?)
- *Module VIII:* Community Relations in the Context of Culture and Multicultural Law Enforcement (Right or wrong, loud or quiet—and who is to say?)
- *Module IX:* Conflict Management and Community Control (Civil disorder or freedom of expression—who draws the line?)
- *Module X:* The Media Link (Guilty until proven innocent and then still guilty!)

The 10 modules are proposed here in the most generic manner, and police academies can modify and customize the modules based on available time and resources. However, the basic formula must be preserved. The current module lasts for 10 weeks, with 10 hours of instruction weekly, 2 hours daily. It is a module of 100 hours and is designed as such to provide for a counter to over 100 hours of training devoted, on average, to use of force/firearms/ self-defense training in an average basic training academy. The basic formula has to do with the overall number of training hours offered by a given academy. If the academy devotes only 50 hours to use of force-related themes, the OCP can be modified to 50 hours as well. If it offers 150 hours of traditional skills, the number of hours devoted to OCP has to be increased as well.

SUMMARY

The chapter provided an overview of the development of the community-oriented policing philosophy as well as the practical side of implementation of the theory through training. Historical developments, traditions, biases, and misconceptions accompanied the various experiments in the implementation of COP. In the first year of the twenty-first century, almost two decades after the first clearly defined tenets of the philosophy were put on paper, the majority of police forces still struggle, while trying to develop a successful training concept that will reorient officers toward this new viewpoint in a practical and effective manner.

The chapter offers a model for community-oriented training to be implemented during the basic academy training. The success of this model is based on the belief that only by providing some balance of new skills against traditional skills can we expect a real shift, both conceptual and practical, in the implementation of practical community-oriented policing training.

REFERENCES

Allport, G. W. (1954). *The Nature of Prejudice.* Cambridge, MA: Addison-Wesley.

Anonymous (1996). Proposals for regional community policing institutes sought. *Management and Training Digest*, December, p. 7.

Anonymous (1999). Police training in the 21st century. *FBI Law Enforcement Bulletin, 68*, 16–19.

Anonymous (2000). Examining police behavior under Nazi rule offers contemporary lessons in moral responsibilities and civil liberties. *Community Policing Exchange, Phase IV, 30*, 3.

Aragon, A., and Adams, R. E. (1997). Community-oriented policing: success insurance strategies. *FBI Law Enforcement Bulletin, 66*, 8–10, 15–16.

Bloch, P. B., and Specht, D. (1973). *Neighborhood Team Policing.* Washington, DC: U.S. Department of Justice.

Bratton, W., and Knobler, P. (1998). *Turnaround: How America's Top Cop Reversed the Crime Epidemic.* New York: Random House.

Breci, M. G. (1997). The transition to community policing: the department's role in upgrading officers' skills. *Policing: International Journal of Police Strategy and Management, 20*(4), 766–776.

Bureau of Justice Assistance (1994). *Neighborhood Oriented Policing in Rural Communities: A Program Planning Guide* (Monograph), August. Washington, DC: U.S. Department of Justice.

California Attorney General's Office (1992). *Community Oriented Policing and Problem Solving (COPPS) Definition and Principles.* Sacramento, CA.

Codish, K. D. (1996). Putting a sacred cow out to pasture. *Police Chief, 63,* 40–44.

Escobedo v. Illinois, 378 U.S. 478 (1964).

Gay, W. G., Day, T. M., and Woodward, J. P. (1977). *Neighborhood Team Policing: Phase I Summary Report.* Washington, DC: U.S. Government Printing Office.

Gideon v. Wainwright, 372 U.S. 355 (1963).

Goldstein, H. (1990). Problem-oriented policing. New York: McGraw-Hill.

Greenwood, P. W., Chaiken, J., and Petersilia, J. (1977). *The Criminal Investigation Process.* Lexington, MA: DC Heath.

Harvey, W. (1999). FTOs: silent support team for COP. *Police, 23,* October, pp. 38–40.

Huber, G. K., Miller, C., and Glick, W. (1993). Understanding and predicting organizational change. In G. Huber and W. Glick (eds.), *Organizational Change and Redesign: Ideas and Insights for Improving Performance.* New York: Oxford University Press.

Kelling, G. L., and Coles, C. M. (1996). *Fixing Broken Windows: Restoring Order and Reducing Crime in our Communities.* New York: Free Press.

Kelling, G. L., and Moore, M. H. (1988). The evolving strategy of the police. In J. R. Greene and S. D. Mastrofski (eds.), *Community Policing: Rhetoric or Reality.* New York: Praeger, pp. 3–25.

Kelling, G. L., Pate, T., Dieckman, D., and Brown, C. E. (1974). *The Kansas City Preventive Patrol Experiment.* Washington, DC: Police Foundation.

Kerner Commission (1968). *Report of the National Advisory Commission on Civil Disorders.* Washington, DC: U.S. Government Printing Office.

Leibernan, G. F. (1972). Cop on the block: a new anti-crime plan is tested in Brooklyn. *New York Times,* December 10, p. 172.

Mapp v. Ohio, 367 U.S. 643 (1961).

McElroy, J. E., Cosgrove, C. A., and Sadd, S. (1988). *CPOP: The Research—An Evaluative Study of the New York City Community Patrol Officer Program.* New York: Vera Institute of Justice.

McLaughlin, V., and Donahue, M. E. (1995). Training for community oriented policing. In P. C. Kratcowski and D. Dukes (eds.), *Issues in Community Policing.* Highland Heights, KY: Anderson Publishing and ACJS, pp. 125–138.

Miranda v. Arizona, 384 U.S. 436 (1966).

Moore, M. H., Trojanowicz, R. C., and Kelling, G. L. (1988). *Crime and Policing.* Washington, DC: U.S. Department of Justice.

National Advisory Commission on Criminal Justice Standards and Goals (1973). *Police.* Washington, DC: U.S. Government Printing Office.

National Advisory Commission on the Causes and Prevention of Violence (1969). *To Establish Justice, to Insure Domestic Tranquility.* Washington, DC: Government Printing Office.

Office of Justice Programs. Retrieved November 5, 2000 from the World Wide Web: *http://www.nara.gov/guide/rg423.html.*

Oliver, W. M. (1998). *Community-Oriented Policing: A Systematic Approach to Policing.* Upper Saddle River, NJ: Prentice Hall.

Oliver, W. M. (2000). *Community Policing: Classical Readings.* Upper Saddle River, NJ: Prentice Hall.

Pate, A. M., Wycoff, M. A., Skogan, W. G., and Sherman, L. W. (1986). *Reducing Fear of Crime in Houston and Newark: A Summary Report.* Washington, DC: Police Foundation.

Peak, K. J., and Glensor, R. W. (1999). *Community Policing and Problem Solving: Strategies and Practices.* Upper Saddle River, NJ: Prentice Hall.

Police Executive Research Forum. Retrieved November 17, 2000 from the World Wide Web: *http://www.policeforum.org.*

Police Foundation (1981). *The Newark Foot Patrol Experiment.* Washington, DC: Police Foundation.

Police Foundation. Retrieved October 23, 2000 from the World Wide Web: *http://www.policefoundation.org/docs/history.html.*

President's Commission on Campus Unrest (1970). *Report of the President's Commission on Campus Unrest.* Washington, DC: U.S. Government Printing Office.

President's Commission on Law Enforcement and the Administration of Justice (1967). *The Challenge of Crime in a Free Society.* Washington, DC: U.S. Government Printing Office.

Riley, J. (1999). Community policing: utilizing the knowledge of organizational personnel. *Policing: International Journal of Police Strategies and Management, 22,* 618–632.

Rosenbaum, D.P., and Lurigio, A. (1994). An inside look at community policing reform: definitions, organizational changes, and evaluation findings. *Crime and Delinquency, 40,* pp. 354–370.

Rosenbaum, D.P., Yeh, S., and Wilkinson, D. L. (1994). Impact on community policing on police personnel: a quasi-experimental test. *Crime and Delinquency, 40*(3), July, pp. 331–353.

Skogan, W. G., and Hartnett, S. M. (1997). *Community Policing: Chicago Style.* New York: Oxford University Press.

Spelman, W., and Brown, D. K. (1981). *Calling the Police: Citizen Reporting of Serious Crime.* Washington, DC: Police Executive Research Forum.

Spelman, W., and Eck, J. E. (1987). Newport News tests problem-oriented policing. *Research in Action,* January/February. Washington, DC: National Institute of Justice.

Spelman, W., and Eck, J. E. (1987). Problem-oriented policing. National Institute of Justice: *Research in Action.* Washington, DC: National Institute of Justice, January/February.

Swanson, C. R., Territo, L., and Taylor, R. W. (1998). *Police Administration: Structures, Processes, and Behavior,* 4th ed. Upper Saddle River, NJ: Prentice Hall.

Trojanowicz, R. C. (1982). *An Evaluation of the Neighborhood Foot Patrol Program in Flint, Michigan.* East Lansing, MI: Michigan State University.

Trojanowicz, R.. Kappeler, V. E., Gaines, L. K., and Bucquueroux, B. (1998). *Community Policing: A Contemporary Perspective,* 2nd ed. Cincinnati, OH: Anderson Publishing.

Weisburd, D., and McElroy, J. E. (1988). Enacting the CPO Role: findings from the New York City pilot program in community policing. In J. R. Greene and S. D. Mastrofski (eds.), *Community Policing: Rhetoric or Reality,* New York: Praeger, pp. 89–101.

Williams, H., and Murphy, P. V. (1967). President's Commission on Law Enforcement and the Administration of Justice.

Williams, H., and Murphy, P. V. (1990). The evolving strategy of police: a minority view. *NIJ: Perspectives on Policing, 13,* 11–12.

Wilson, J. Q., and Kelling, G. L. (1982). Broken windows: the police and neighborhood safety. *Atlantic Monthly, 249,* 29–38.

Zhao, J., Thurman, Q. C., and Lovrich, N. P. (1995). Community oriented policing across the U.S.: facilitators and impediment to implementation. *American Journal of Police, 16,* 11–28.

Zhao, J., Thurman, Q., and Simon, C. (1999). Innovation in policing organizations: a national study. In D. J. Kennedy and R. P. McNamara (eds.), *Police and Policing: Contemporary Issues.* Westport, CT: Praeger.

MULTICULTURAL LAW ENFORCEMENT

INTRODUCTION

The final outcome of the Los Angeles riots after the Rodney King verdict has to be evaluated not only in terms of the social and financial cost U.S. society had to absorb and deal with, but also by looking at its impact on law enforcement training in the United States. The concept of multicultural training for police became a household word in many larger police departments across the country, and police scholars and consultants began to write about and advocate for diversity training to be included during the academy and on-the-job training. A number of states made it mandatory for police officers to undergo diversity training in order to be certified or recertified as sworn police officers. From describing a life-long approach to diversity training to as little as a few hours of training, the chapter culminates with a training module proposed by the author referred to as multi-cultural close contact training, the MCC model.

FOUNDATION OF DIVERSITY TRAINING

The idea of diversity training for law enforcement is not new; it can be traced to the early 1940s, when the International City Management Association (ICMA) published its first manual for race relations, entitled *The Police and Minority Groups*. Other professional organizations were also involved in the preparation of training manuals and participated in some race-related training for police officers. Academics, such as Joseph D. Lohman from the University of Chicago, published a widely used training text. The Governor of California, Earl Warren, enthusiastically supported the newly emerging programs (Barlow and Barlow, 2000).

It is interesting to note that most programs dealing with race-related issues developed in the 1940s are quite similar to programs developed in the 1960s and the cultural diversity training modules introduced in the 1990s. The strategy was based on three main concepts (Walker, 1992):

1. Race relations training for rank-and-file police officers
2. Development of formal contacts with black leaders, along with the hiring of additional African-American police officers
3. Establishing guidelines for handling civil disorder.

Despite the fact that numerous police departments and academies experimented for almost six decades with various forms of sensitivity/multicultural/race-related training modules, it appears that the situation today does not differ much from the situation in the 1940s as far as police–minority relations are concerned. Training concepts, manuals, and programs do exist, and are implemented by police departments with varying degrees of success. However, the tension between "them" (the police force) on the one hand and "us" (various minority groups) appear to be even more pervasive than ever before.

In this chapter we provide the reader with a sample of ideas, some suggested by scholars and trainers on a rather theoretical plateau, others actually designed and implemented by police departments around United States and in other countries. Although we discuss training comparatively elsewhere, it is relevant here to introduce some international approaches to the topic of diversity. The primary reason for inclusion of the international perspective in this "practical" chapter is the desire to dispel the notion of problems and conflicts between police and the public in the United States as being based primarily on the tensions between the police and certain minorities. By bringing ideas from other countries, the author hopes to emphasize one of the major themes of this book, the idea that police problems are related directly to the role that the police have had, traditionally and historically, in any society, and that this role continues to be cultivated through the training that police officers receive, despite the rhetoric of community-oriented policing. Research indicates that 90 percent of major civil disorders in the United States have resulted from police–citizen conflicts, many of which could have been avoided (Fisher-Stewart, 1994).

In 1994, the highly respected International City and County Management Association Report included in its volume an article by Fisher-Stewart, a retired captain in the Metropolitan Police Department in Washington, DC. In her article, Fisher-Stewart presented suggestions for integrating diversity training into ongoing learning at all levels of the police department. The article included a model for training police executives to examine and modify policies, practices, and procedures dealing with managing issues of diversity within the department. The ideas presented in the article combine community policing training modules with diversity training. Although both topics are obviously related, the training concepts are too complex to be conveyed in a proper manner in one training module.

For example, a scenario is presented to provide an illustration of how to handle a municipal/criminal code violation. A second scenario illustrates dealing with investigative techniques while dealing with people of various cultures. While both scenarios appear to contain valid information about the particularities of a given culture, such as offering a bribe to a police officer (based on the fact that in some Latin American countries the way to do business with any public official, particularly police, is to offer money) and avoiding eye contact during questioning (indicative of respect in certain cultures), they miss the critical point of going beyond the awareness of certain culture-related particularities.

Fisher-Stewart outlines a training module for police executives consisting of the following seven segments (1994, p. 10):

- *Part I:* definition of community-oriented policing and why understanding diversity is important.
- *Part II:* managing diversity
- *Part III:* the meaning of culture
- *Part IV:* institutionalizing community-oriented policing
- *Part V:* marketing the new values
- *Part VI:* leadership for community-oriented policing
- *Part VII:* strategies for evaluating the success of community-oriented policing and diversity management

This seven-part program, one of the earliest serious attempts to provide diversity training for police departments, although it touches on very valuable themes, appears to try to convey too much, too early, in a too complex manner. The concepts of community-oriented policing, woven into concepts of diversity, leadership, and evaluation, are too much to be processed by any audience, even when conveyed in a 40-hour introductory course spread over several days or weeks with follow-up assignments and biweekly 2- or 4-hour workgroup sessions that continue until retirement (as Fisher-Stewart advocated). She should be commended for this attempt to introduce complex training themes and advocating for "never-ending" training sessions with mixed ranks and ongoing evaluations. However, to a certain degree, the mix of themes invalidates the messages that are being conveyed. What she fails to recognize is the fact that as mentioned in earlier chapters, basic academy training does not prepare the audience for diversity training, but to the contrary, by concentrating on the traditional use-of-force curriculum, it generates quasimilitary "understanding" of the surrounding environment. An academy that emphasizes the importance of physical fitness of its instructors over their intellectual abilities (or at a minimum assigns the same degree of importance to physical and intellectual capabilities) does not produce officers easily amenable to change by being exposed to even the most extensive diversity training.

Police departments that disqualify potential candidates for police work if a candidate cannot perform a certain number of push-ups during a defined time frame, and disqualify others based on some of their physical attributes and do not even allow them to take the remaining portions of the tests, do not generate the "right" audience. To provide a successful framework for diversity training, a department or academy has to clearly separate multicultural training modules from any related themes. Only by concentrating intensely on the issues of multi-culturalism can an agency hope to deliver an effective training module.

In 1995, Shusta et al. published a very much needed textbook entitled *Multicultural Law Enforcement*. Besides being very well written and containing a volume of information regarding race-related issues, the book includes a chapter that deals specifically with the preparation and implementation of cultural awareness training for law enforcement. The authors provide sample agency courses from four agencies, ranging in length from a three-day course offered by the NYPD to a 48-hour course designed by the Los Angeles Sheriff's Department, with Concord, California (16 hours) and Metro–Dade, Florida (24 hours) in the middle. The entire chapter, devoted to training, is replete with extremely useful tips and innovative ideas on how to design and successfully implement diversity training, from training concepts, through selection of the trainers, training delivery methods, and suggestions for themes to be included in the modules.

It appears to the present author (who uses this textbook in her police training courses, and whose MCC model's training modules are based on many of the book's subheadings) that the only problem is the lack of emphasis on the fact that for small police departments with 50 or fewer sworn officers, which represent the majority of police forces in the United States, the ideas and tips suggested are too elusive. Larger cities, with in-house academies, can use the ideas and, with some good will, implement most of them. Small police departments generally cannot afford or are not willing to allocate the needed resources.

In 1997, Gould published an article, based on a study that evaluated the reactions of 151 police officers, both cadets and experienced officers, to a course in cultural diversity. The police officers were in various stages of their careers, from beginning cadets through officers with up to 18 years of experience. The diversity course offered required 16 hours of instruction over a two-day period and took place at a police training facility in a southern state. This course was incorporated into the academy training agenda, which each new police officer is required to attend. The proposition of the study was based on the assumption that experienced officers' ability to listen and truly hear new ideas may be so diminished that teaching cultural diversity issues is lost on them.

The findings of the study indicated that compared with the more experienced officers, cadets were generally more positive about and receptive to the course, the information contained in the course, and the usefulness of the course to police officers. Experienced officers felt that the course was a waste of time, and their criticism was summarized in five points (Gould, 1997, p. 351):

1. A feeling that the community did not understand or appreciate what the officers were trying to accomplish
2. A feeling that most police administrators and many supervisors had lost touch with the reality of policing as officers face it today

3. A feeling that many police administrators and community politicians were looking for a quick-and-dirty scapegoat, often blaming police officers for things over which they have no control

4. A feeling that the "rules of the street" are far too often weighted against the police

5. A feeling that there is a divergence between what is being taught in the course and what society actually asks a police officer to do

Gould has suggested that some policies be considered based on the findings of his study. He stresses the following points as they relate to future diversity training designs (Gould, 1997, pp. 354–355):

1. One should remember that teaching diversity also means the "un-teaching" of some already existing culturally intensive attitudes and behaviors.

2. A change in behavior will not generally result from sitting through one cultural diversity course.

3. For the training to have its greatest effect, it should be tailored to meet the needs of officers as well as the community.

4. The training of experienced officers should include the training of administrators in the same classroom setting.

5. Experienced officers' training should include more time for venting of frustrations centering on the cultural diversity training.

6. Cultural diversity training should begin early in an officer's career.

7. The training should be reinforced throughout an officer's career.

8. The training should be aimed toward explanation and discovery concerning cultural differences rather than appearing to place blame for police–community conflict on the individual officer.

Also in 1997, the Charlotte–Mecklenburg Police Department, in collaboration with the local community, introduced an innovative diversity training module. Since the early 1980s, the Community Relations Committee in Charlotte has provided training in mediation and conflict management, fair housing, sexual harassment, race relations and diversity to volunteers, community and professional groups, schools, police officers, and city and county employees.

During the summer of 1997, committee staff became involved in a collaborative effort with the Charlotte–Mecklenburg Police Department to develop a diversity training module for all personnel of the department, civilian and sworn. This partnership has resulted in a training process that was designed to engage police personnel in a constructive dialogue of the issue of diversity within the force and the Charlotte–Mecklenburg community. The training style was designed to engage all participants in the process and provide a safe environment where differences can be discussed without fear of reprisal. The training was implemented initially with two pilot groups of police officers with very positive results (Ratchford, 1998).

Today, the one-day training module offered by CMPD's in-house training academy consists of the following modules.

Module I: Ground Rules

A. It is OK to disagree.

B. Be honest and respectful.

C. There are no reprisals.

D. Keep what is said in the room.

E. Be an active listener.

F. Try to be open-minded.

G. Put yourself in the speaker's shoes.

H. Make no personal attacks.

I. One person speaks at a time.

Icebreaker

Objective: To get acquainted and set a tone for the training. Icebreakers may or may not be linked to the training subject matter. Icebreakers provide an opportunity for participants to relax and feel comfortable about the training. (Time required: 5–15 minutes, depending on the icebreaker)

Module II: What Is in the Room?

Objective: To illustrate the many forms of diversity in the room, to gauge what groups the participants are comfortable with, to help the participants become more familiar with one another, and to determine common ground. (Time required: 30–45 minutes)

Module III: Hopes and Hesitations

Objective: To identify any concerns the participants may have about participating in diversity training and address those concerns. To identify any hopes that participants may have regarding their participation. At the end of the session, participants are given an opportunity to discuss whether their hesitations were real or well founded, and whether their hopes were met. (Time required: 10–15 minutes first part, 30 minutes second part)

Module IV: Definitions Exercise

Objective: To get an understanding of what discrimination, racism, stereotype, prejudice, and bigotry mean to the participants in the session; and to help them to see that their definitions are shaped by their life experiences and individual perceptions. This does not mean that one definition is right while another is wrong but that they are different depending on the individual. During a discussion, note whether different perspectives may be shaped by race/ethnicity. (Time required: 45 minutes)

Module V: Video/True Colors
Objective: Many in the community question the extent to which racial/ethnic discrimination still exists in our everyday lives. Recent survey results of race relations polls were published which show that blacks and whites have very different views on the state of race relations. This video will show that discrimination does exist and that it is often very subtle and sophisticated. Because of the subtle ways it is now practiced, discrimination often goes unnoticed by victims of this unfair treatment. (Time required: 45 minutes)

Module VI: Telling Our Stories/Time Line
Objective: Decided and defined by the trainer

Module VII: Video/ Brown Eyes—Blue Eyes
Objective: To illustrate the negative effects of internalized oppression and how it works to destroy self-esteem and a person's sense of worth; to illustrate the benefits of positive reinforcement on self-esteem and one's sense of worth. During the discussion phase of the exercise the facilitator should have participants discuss the implications for race and ethnicity. (Time required: 1 hour)

Module VIII: Wrap-Up/Revisit Hopes and Hesitations
Objective: Conclude the session by revisiting hopes and hesitations. Open floor for discussion, for participants to share their final thoughts on diversity, the police department, and the training. Administer class evaluation of the training. (Time required: 30 minutes)

The time devoted to each segment of training paired with or contrasted against the importance of any given theme will determine the true effectiveness of this module.

The city of Cincinnati, Ohio, and its in-house training academy deserve a few words of praise for incorporating an interesting approach to its sensitivity training module, which is also incorporated into its basic recruit training curriculum. The Ohio Peace Officer Training Commission requires only 474 hours of training; the Cincinnati Police Department training almost doubles the requirements, to a total of 880 hours of training. The cultural sensitivity modules are introduced during the eighth week of the academy and delivered in 9-hour blocks with a break for lunch. The two instructors responsible for the module are civilians who have received a special certification to teach at the police academy. The state POST mandates that all instructors at the academy be sworn police officers; however, an exception has been made, and the two civilians received special certifications to be able to deliver the training. They bring in guest lecturers from the community, who represent diverse groups of the populations: gays and lesbians, ethnic minorities, disabled people, and so on (Gardner, 2000).

Some in-house academies took a slightly different approach to diversity training by offering the instruction under separate blocks, "hidden" under disguised headings. For example, while examining the basic training offered by the

New Orleans Police Department, it is surprising to find that they devote only 1 hour to sensitivity training. However, upon a more thorough review of the curriculum, some closely related themes add up to a more expanded approach. For example, topics such as "Minority Relations," with 2 hours of instruction, or " The Gay Community," with 3 hours, can certainly be considered as part of the sensitivity training, even though they are oddly distributed. (Actually, the gay community theme has a heading of sensitivity training as well, but it is difficult to establish why it is separated from the 1-hour instruction module entitled "Sensitivity Training.")

The Los Angeles Sheriff Department's in-house training academy offers a total of 919 hours of instruction, while the California POST mandates only 664 hours. In the basic academy training curriculum for 1995, the last block of instruction is devoted to cultural diversity training, which is delivered in 26 hours and includes the following topics:

- Orejudice discrimination
- Mainstream America
- Hispanic issues
- African-American issues
- Asian/Pacific Islander issues
- Gay and lesbian issues
- Sexual harassment
- Hate crimes

Curiously enough, the LASD academy schedule revised in 1999 lasts for 18 weeks and does not include all cultural diversity modules that were listed in 1995; instead, the leading theme appears to be leadership training. Leadership training is delivered, starting with the first week of instruction, for 4 hours, supplemented by "Tactical Communications" for 2 hours (*tactical communication* is the new term for courtesy and good manners).

An additional hour of tactical communications, 3 hours of leadership, and 4 hours of hate crimes are offered during the second week. During the fourth week, 2 more hours of leadership, 1 hour of leadership during the fifth, seventh, and eleventh weeks, 2 hours during the twelfth week, and 8 hours during the last week: all together, 23 hours of leadership training, which apparently replaced sensitivity training.

The academy is one of the very few in the country that actually delivers intensive leadership training, during the duration of the basic training, and does it in an incremental mode, again unrepresentative of basic instruction elsewhere. Nevertheless, it is slightly puzzling why cultural sensitivity was replaced by leadership modules instead of being supplemented with this theme. In 1999, the NYPD came under harsh attack for the fatal shooting of Amadou Diallo. Calls for more sensitivity training were heard and addressed. Cultural sensitivity training for rookie police officers in New York City was created well before the shooting but was enhanced after the shooting. The training program, called "Streetwise: Language, Culture, and Police Work in New York City," was originally funded in 1998 by the Justice Department. The curriculum was created under the auspices of the New York State Regional Community Policing Institute (RCPI).

More than 2500 NYPD rookies and in-service personnel underwent one-day training between April and May 1999. In addition to providing a card that can be inserted into the officer's books containing basic expressions in Spanish, Haitian Creole, Russian, and Mandarin Chinese, the curriculum includes a series of handouts that explore the cultural dimensions of each group and various city neighborhoods, along with an audio tape.

A fifth component, the African-Caribbean/American Experience, was developed by the John Jay College of Criminal Justice, one of four partners that make up the RCPI, along with the New York City Police Department, the State Bureau of Municipal Police, and the Citizens Committee for New York City.

The curriculum also discusses responses common to an individual group which could help defuse a potentially violent confrontation. In developing the program, the RCPI used focus groups in the community and within the NYPD and participated in ride-alongs with officers (Anon., 1999). The one-day training is offered as an addition to the in-house diversity training developed by the NYPD.

The idea of handouts included in the human diversity portion of the basic recruit training is also utilized by the regional training academy in St. Petersburg, Florida. The handouts contain numerous definitions and terms relevant to the training. Following is a sample of themes from the handouts, certainly an extremely valuable instructional and learning tool.

- *Handout 1-5:* Human Diversity Definition
- *Handout 3-5:* Cultural Attributes versus Stereotypes
- *Handout 3-9:* Six Guidelines for Bridging Barriers
- *Handout 4-6:* Defining Racism and Sexism

INTERNATIONAL EXAMPLES OF DIVERSITY TRAINING

To provide some additional background for introduction of the MCC model, a number of international experiments with diversity training are reviewed below. What appears to be quite clear from this review is the fact that diversity training seems be introduced whenever a given country experiences an influx of foreigners, legal or illegal immigrants. What seems to be paradoxical is the fact that more than any other county, the United States is a country of immigrants but does not include this ongoing theme in its police training. If anything, the constant influx of new immigrants to this country should be the best indicator of and testimony to the dire need for inclusion of mandatory and extensive cultural awareness training in all basic police training curricula.

Belgium: Building Awareness of Immigrant Issues

In March 1989, the Belgian government appointed a Royal Commissioner for Immigrant Policy for a term of four years. One of the recommendations of the royal commission culminated in the introduction of a 25-hour training program for policemen and gendarmes from various towns and municipalities, allotted as follows.

- *Basic information*–3 × 3.5 hours: history and development of immigration, current situation, demography, cultural and religious origins, young immigrants, etc.
- *Practical information*—2 × 3.5 hours: discussion on practical experience, breakdown of communication, conflict situations, etc.
- *A day of contact in the field*–7.5 hours: encounter between the participants and the local immigrant population in a community setting (youth center, mosque, women's club, etc.)

Denmark: Police Activities toward and among Aliens

The basis for the establishment of this specialized course was the adoption of a new Aliens Act, which took effect in 1984. A one-week course was designed for the Danish Police School and Staff College and affected other courses provided by the police school. The basic training program underwent changes in the subjects of civil law, psychology, and police theory. Within individual police districts the issue was discussed within the framework of the SSP (Schools, Social Service, and Police) operation scheme. The selection of subjects covered a broad perspective, with an emphasis on information on migrants, especially refugees. Subjects chosen were:

1. Introduction to the Aliens Act
2. Meeting of cultures (course conducted by an ethnographer and a psychologist)
3. Integration (course members work with theories and experiences concerning dissimilar cultures and their integration, conducted by an ethnographer)
4. Danish civilization (course conducted by an ethnographer)
5. International "hotbeds" (subject taught by a Foreign Office official, academic, or journalist)

Finland: Foreigners

The basic police training concerning "foreigners" has traditionally included two elements: legal ramifications and ethnic relations. However, the new Aliens Act issued at the beginning of 1991 led to a reorganization of training concerning foreigners.

Police School

Police training in Finland lasts for 12 months, during which students are taught primary skills and knowledge of police work. After having completed at least one year of practical training as constables, students return for the second, more extensive basic course, lasting five months. In both courses, lectures on the Aliens Act and its practical effects on police work are given by a superintendent involved in work with foreigners. Furthermore, additional lectures are delivered on minority cultures and some psychological concepts, such as xenophobia, attitudes, and stereotypes.

Police College

The Police College offers two different types of training concerning foreigners. The police officers of higher rank are given regular lessons, which included social and psychological topics, together with a two-day seminar about issues concerning foreigners. Furthermore, for those officers who are in regular contact with foreigners, special two-week courses are available.

France: Toward a Multicultural Society

The Centre Nationale d'Études et de Formation, responsible principally for in-service training for police officers, has introduced three major training initiatives in the area of police–minority relations. In 1983 the National Police Charter set out 12 goals, two of which were concerned with police officer training in their relations with immigrants: goal 5, fostering better understanding between police officers and the public, and goal 7, enabling police officers to analyze social factors. In 1989, in the wake of an assessment of the training policy, two more goals were added: improving public relations and promoting initiatives to improve understanding of social phenomena.

Between 1988 and 1992, three summer school courses were introduced to police officers.

1. *Teenagers and Immigration:* provided information about young immigrants
2. *Teenagers in France and North Africa:* emphasized common backgrounds and features of all young people
3. *The Integration or Exclusion of Young People in France and Europe:* dealt with the processes of integration and exclusion peculiar to young people

The summer schools were week-long programs, prepared by universities and researchers, including afternoon working groups and field visits. In addition, a course entitled "Cultures and Cultural Differences" was designed and has been offered as three two-day modules once a month. Each module concentrated on a single immigrant community.

The methodology of instruction was slightly nontraditional. At the summer schools and during the training course the real-life experience of the trainees was exploited. Instead of approaching the topic from a more academic perspective of theories, it was decided to tackle the issues on the basis of the practical experience of those involved in the training (Council of Europe, 1998).

To culminate the review of sensitivity training concepts in the United States and around the world, the author would be remiss not to mention the "Human Dignity and the Police" course developed by the John Jay College of Criminal Justice. The course, delivered during a three-day session, was designed to achieve experiential learning by using nontraditional teaching and learning methods. The idea for the course was first "planted" in 1994 and has developed over time, with aims to achieve a broad range of objectives (Lynch, 1999, pp. 6–7):

1. To have the participants so imbued with the innateness of human dignity that it affects their personal and professional attitudes and behavior.
2. To encourage participants to ask important, soul-searching questions.
3. To enhance participants' own human dignity.
4. To make the course experience exciting and dramatic.
5. To make the course experience free-flowing so that as in all successful enterprises, the outcome appears to be natural.
6. To draw on the talents and experiences of a diverse group of scholars and law enforcement professionals.
7. To bear in mind that the course is never a completed work. It is a continuous process, an extraordinary opportunity, an unfolding mystery, and an intelligent and compelling story, which is modified and adapted to by each participant, group, country, and culture.

The course is composed of the following sessions:

  I. Introduction/Orientation (½ hour)
 II. Human Dignity: Working Definitions; Historical Perspective (1½ hours)
III. Human Dignity: Violation of Human Dignity by an Authority Figure (2 hours)
IV. Human Dignity: Abuse of Power by Police (1½ hours)
 V. Human Dignity: Peer Pressure in Police Organizations (1½ hours)
VI. Police and Society's Outcasts (1½ hours)
VII. The People's Perception of Their Police (1 hour)
VIII. Listening to Community Concerns (1 hour)
 IX. Human Dignity, Integrity, and Professional Responsibility (1 hour)
  X. Human Dignity: Police Codes of Conduct, Practical Applications (2 hours)

The course is of extreme value, especially to the newly transformed police forces with traditions of oppressive behavior under various totalitarian regimes, but should certainly be considered as a value-added contribution to any basic police academy in the United States.

MULTICULTURAL CLOSE-CONTACT TRAINING MODEL

Conceptual Framework

The idea behind the development of the multicultural close-contact (MCC) training model was born in 1996, when this author was asked to deliver a one-day diversity training module to Jersey City, New Jersey police recruits during the final week of their training. The day scheduled for sensitivity training coincided with the first day during which recruits received their new uniforms, adorned with all the military insignia of power, the most important of which was their new gun. The training was delivered in an old facility, a bus hanger, with no air conditioning (in the middle of a very hot June) and a lot of noise from outside. The importance of facilities and other environmental factors on effective delivery training cannot be overstated and is discussed in Chapter 15. However, the timing of any training cannot be underestimated, ignored, or dismissed by anybody who claims to be serious about police training. Diversity training cannot be delivered effectively to people who already have the feeling of being different—by wearing the uniform and gun, and projecting power—at least externally to the public, if not also to themselves. To be realistic, effective diversity training must be incorporated into basic academy training not during the last weeks of training but from the very start, and has to last through the duration of the academy.

The proposed MCC model is shorter than the length of an average academy but was designed to be incorporated easily even into the shortest training schedules. For academies with longer curricula it is advised to include a segment of the MCC model every other week, to ensure the effectiveness of the close-contact interaction. The premise of this module/course is its mandatory inclusion in the basic academy training for police officers. The premise was developed based on the equal status contact hypothesis, which emphasizes the need for intense intergroup contact as a remedy for prejudice, misunderstanding, and hostile race and ethnic relations.

The first factor in the four-pronged hypothesis is the *equal-status concept*. According to this idea, intergroup contact is likely to reduce prejudice only when the people involved are equal in status. When interaction fails to meet this standard, it is more likely to sustain or even intensify prejudice, since it mirrors the power differentials and inequalities of the larger society. Therefore, by customizing the equal-status concept to the MCC module, we cannot allow for mixing of ranks during training sessions. Therefore:

- *Principle 1*. There should be no mixing of ranks (contradictory to the equal-status hypothesis).

The second factor is the *intensive interaction concept*, which stresses that contact must be more than superficial. The situation must last for a significant length of time, and the participants must be fully involved. A more-than-superficial environment can be created only during basic academy training, since even the most extensive on-the-job diversity training misses the goal of intense interaction. Therefore:

> ▪ *Principle 2.* Intensity can be achieved only over a lengthy period of time, not in modules lasting from one to three days (contradictory to the intense interaction hypothesis).

The third factor is *noncompetitive relations.* For contact to reduce prejudice, it must occur in an atmosphere free of threat or competition between groups. On-the-job training modules are, by definition, given to personnel who are in a certain state of competition, either for a promotion to a higher rank or for a different position. Therefore, the nature of a given work environment already precludes noncompetitive relations. Recruits during basic academy training, especially when this training module is introduced from the very start, are still relatively free of the competitive group "bug." Therefore:

> ▪ *Principle 3.* The training module has to be introduced from the very start of the training session, to eliminate the creation of competitive group relations.

The fourth factor is *cooperative tasks.* Intergroup contact will be most likely to reduce prejudice if it goes beyond the mere absence of competition. The most effective contact situations are created when members of all groups are required to cooperate in completing a task that could not be accomplished by the efforts of one group alone or by people acting autonomously. When people are bound together to accomplish a clearly defined task, they are much more likely to regard one another as individuals and not as stereotypical representatives of their groups. Therefore:

> ▪ *Principle 4.* A cooperative task has to be assigned from week 1 of the academy, a task that requires input from all the participants.

Instructional Modules

The instructional modules are based on the required textbook by Shusta et al. (1995).

Module Description

This course explores the pervasive influence of culture, race, and ethnicity on daily encounters, contacts, and interaction between police officers and civilian employees of police organizations and other community members. The emphasis is on cross-cultural contact, the need for awareness, understanding of cultural differences, and respect for those of different backgrounds. The nature and concepts of the police role in the context of an increasingly large multicultural society are discussed.

Goals of the Module

American society is still preoccupied with race, ethnicity, and diverse cultural orientations. These critical issues divide and define our society. Law enforcement, in its essence, can be complex, painful, and problematic, regardless of the multicultural dimensions. The goal of this module is to analyze the concepts of racial, ethnic, and cultural stereotypes and to evaluate the impact of prejudice on police professionalism. Suggestions for improving law enforcement in multicultural communities are the desired outcomes of the course.

Instructional Procedures

1. *Lectures.* The specific goals of this module are aimed to be met by utilizing the following techniques:

   (a)   During the first meeting, various racial/ethnic/racial groups are listed on the board. This diversity represents the students' (recruits') backgrounds. Each student is asked to pick up one group (not his own) and describe it in a few sentences. The answers are written on plain cards, with no identification, supplied by the instructor, to assure the anonymity of the answers. Each student gets two cards. The first card is filled out during the first class meeting and submitted to the instructor; the second is filled out toward the end of the semester. Students are asked to write about the same group and mark their cards in a way that enables the instructor to match them. Attitudes "before" and "after" being exposed to this course are discussed during the final meetings.

   (b)   To increase students' awareness and understanding of each other, based on their multicultural backgrounds, they are divided into working groups and asked to prepare a short presentation about their colleagues' cultural/racial/ethnic backgrounds. The exercise is based primarily on the information volunteered by one student to another, supplemented by additional research work . The final outcome should demonstrate lack of sufficient knowledge regarding one's own heritage and the implications of that deficiency.

   (c)   To increase the students' effectiveness in roles related to multicultural law enforcement, students are divided into working groups and asked to engage in role-playing, with one group representing law enforcement personnel interacting with various cultural/racial/ethnic groups. Role reversal is utilized as a follow-up to the first exercise.

2. *Videocassettes.* Videocassettes containing programs dealing with racial/ethnic/cultural stereotypes are presented to help students become more cognizant of various perceptions and attitudes and how they shape and influence encounters between law enforcement personnel and civilians.

3. *Guest lecturers.* Guest lecturers with diverse cultural/racial/ethnic backgrounds (law enforcement practitioner and community members) are invited to share their experiences based on encounters in which their race or ethnicity played a major role.

Module Content

Module I: Multicultural Communities: Race, Ethnicity, and Crime
A. Introduction
B. Challenges for Law Enforcement
C. Challenges for the Society
D. Police Knowledge of Cultural Groups
E. Prejudice in Law Enforcement
F. Institutional Behavior
G. Case Studies: Culture and Crime

Module II: The Changing Law Enforcement Agency: Interaction with the Society
A. Introduction: Victims and Offenders
B. Historical Overview: Myths and Realities
C. Changing Workforce
D. Women in Law Enforcement
   - Leading Role in a Pluralistic Society?
   - Archaic Role in a Pluralistic Society?

Module III: Past—Present—and the Past Again?
A. Introduction: Violent Encounters between Police and Minorities
B. History of Profiling
C. Recruitment Strategies
D. Discrimination in Employment Decisions
E. Laws and Court Decisions
F. Retention and Promotion of Women and Minorities
G. Reverse Discrimination Issues
H. Coping with Marginal Status

Module IV: Cultural Awareness or Cultural Ignorance?
A. Introduction: Who Are You?
B. Cultural Awareness: Who Are "They"?
C. Are You "Better"?

D. Coping with Frustration

E. Tolerance versus Prejudice

Module V: Inter- and Intra-Cross-Cultural Communication

A. Introduction: Is It the Language or the Perception?

B. Cross-Cultural Communication in a Law Enforcement Context

C. Language Barriers and Power

D. Attitudes and Power

E. Nonverbal Communication

F. Cross-Gender Communication in Law Enforcement

Module VI: Case Studies

A. African Americans

B. Asian/Pacific Americans

C. Latino/Hispanic Americans

D. Middle Eastern Americans

E. American Indians

Module VII: Hate and Bias Crime: History and Priorities

A. Introduction: The Hate/Bias Crime Problem

B. Historical Perspectives

C. The Impact of Ignorance on Effective Enforcement

D. Law Enforcement Response Strategies

E. Individual Discretion

F. Prioritizing in Law Enforcement

Module VIII: Agitation versus Accommodation

A. Introduction

B. Impact of Officer's Image on Human Behavior

C. Work Habits and Practices

D. "Cosmopolitan" Public Servants

E. Sensitive Peacekeeping

Module IX: Professionalism and Peacekeeping Strategies in a Diverse Society

A. Introduction

B. Leadership in Professionalism

C. Regional/Statewide Cooperation in Law Enforcement

D. Professionalism, Ethics, and Minorities

E. Peacekeeping Strategies in a Multicultural Society

F. Organizational Trends versus Individual Propensities

G. Controlling the Multicultural Law Enforcer

SUMMARY

Hennessy (1997) observed that as the world shrinks and becomes more multicultural, developing the most effective method to train and educate people in law enforcement as to understanding a changing cultural perspective of the citizens they serve will become more critical. Despite the fact that this observation is certainly a valid one, it seems that in the United States we had to struggle with multiculturalism from the inception of this country. It is therefore even more imperative to recognize finally the need to include mandatory and intensive multicultural training in our basic academies. Multiculturalism is basic to our society, as much as use of force is basic to police work. One can only imagine the "hue and cry" that would have been risen by police trainers if use of a force training module had been reduced to a block of instruction lasting between 1 and 8 hours. Why is it, then, that the same hue and cry is not been heard in reference to the multicultural theme?

Bibliography for the Modules

Required Textbook for the Training Module

R. M. Shusta, D. R. Levine, P. R. Harris, and H. Z. Wong (1995). *Multicultural Law Enforcement: Strategies for Peacekeeping in a Diverse Society*. Upper Saddle River, NJ: Prentice Hall.

Additional Recommended Readings

Alpert, G.P., and Dunham, R. C. (1988). *Policing Multiethnic Neighborhoods*. Westport, CT: Greenwood Press.

Felkens, G. T., and Unsinger, P. C. (1992). *Diversity, Affirmative Action and Law Enforcement*. Springfield, IL: Charles C Thomas.

Harris, P. R., and Moran, R. T. (1991/1992). *Managing Cultural Differences*. Houston, TX: Gulf Publishing.

Hendricks, J. E., and Byers, B. (1994). *Multicultural Perspectives in Criminal Justice and Criminology*. Springfield, IL: Charles C Thomas.

Relevant Periodical Sources

Journal of Police Science and Administration
Law and Order
Police
Police Chief
Police Journal
Police Quarterly
Police Studies

REFERENCES

Anonymous. (1999). *Law Enforcement News.* NYPD's "Streetwise" cultural sensitivity training gets renewed impetus. Vol. 25, 153, Iss. 511–512, p. 1.

Barlow, D. E., and Barlow, M. H. (2000). *Police in a Multicultural Society: An American Story.* Prospect Heights, IL: Waveland Press.

Bayley, D., and Mendelsohn, H. (1969). *Minorities and the Police.* New York: Free Press.

Benson, K. (1992). Black and white on blue. *Police,* August, p. 167.

Chin J., Comas-Diaz, L., and Griffith, E. E. H. (1988). *Cross-Cultural Mental Health.* New York: Wiley.

Colrad, A. L. (1992). Foreign languages: a contemporary training requirement. *FBI Law Enforcement Bulletin, 61,* 20–23.

Council of Europe (1998). *Police Training Concerning Migrants and Ethnic Relations.*

Dantzker, M. L. (1995). *Understanding Today's Police.* Upper Saddle River, NJ: Prentice Hall.

Fisher-Stewart, G. (1994). Multicultural training for police. *MIS Report, 26,* 1–18.

Gardner, M. (2001). Personal Communication.

Goldsmith, A. (1990). Taking police culture seriously. *Policing and Society: International Journal of Research and Policy, 1,* 91–114.

Gould, L. A. (1997). Can an old dog be taught new tricks? Teaching cultural diversity to police officers. *Policing, 20,* 339–356.

Hammond, T., and Kleiner, B. (1992). Managing multicultural work environments. *Equal Opportunities International, II.*

Hennessy, S. M. (1997). Multicultural awareness training structure with Arizona police recruits. *Crime and Justice International, 13,* 9–11.

Jamieson, D., and O'Hara, J. (1991). *Managing Workforce 2000: Gaining the Diversity Advantage.* San Francisco: Jossey-Bass.

Johnson, L. (1991). Job strain among police officers: gender comparisons. *Police Studies, 14,* 12–16.

Kennedy, J. F. (1986). *A Nation of Immigrants.* New York: Harper & Row.

Klockars, C., and Mastrofski, S. (1991). *Thinking about Police: Contemporary Readings,* 2nd ed. New York: McGraw-Hill.

Libolt, A. (1991). Bridging the gender gap. *Corrections Today, 53,* December.

Lynch, G. (ed.) (1999). *Human Dignity and the Police: Ethics and Integrity in Police Work.* Springfield, IL: Charles C Thomas.

Marger, M. N. (1991). *Race and Ethnic Relations: American and Global Perspectives,* 2nd ed. Belmont, CA: Wadsworth.

Martin, S. (1990). *The Status of Women in Policing.* Washington, DC: Police Foundation.

Mayhall, P. D. (1994). *Police–Community Relations and the Administration of Justice.* Upper Saddle River, NJ: Prentice Hall.

More, H. (1992). Male-dominated police culture: reducing the gender gap. *Special Topics in Policing.* Cincinnati, OH: Anderson Publishing.

Ogawa, B. (1990). *Color of Justice: Culturally Sensitive Treatment of Crime Victims.* Sacramento, CA: Office of the Governor, State of California.

Pitter, G. E. (1992). Policing cultural celebration. *FBI Law Enforcement Bulletin, 61.*

Pobrebin, M. (1986). The changing role of women: female police officer's occupational problems. *Police Journal, 59.*

Ratchford, W. (1998). Personal Communication.

Sherman, S. (1986). When cultures collide. *California Lawyer, 6.*

Shusta, R. M., Levine, D. R., Harris, P. R., and Wong, H. Z. (1995). *Multicultural Law Enforcement: Strategies for Peacekeeping in a Diverse Society.* Upper Saddle River, NJ: Prentice Hall.

Taft, P. B., Jr. (1982). Policing the new immigrant ghettos. *Police,* July, 10–26.

Takaki, R. (1989). *Strangers from a Different Shore: A History of Asian Americans.* Boston: Little, Brown.

Thibault, E. A. (1995). *Proactive Police Management*, 3rd ed. Upper Saddle River, NJ: Prentice Hall.

Trojanowicz, R., and Carter, D. (1990). The changing face of America. *FBI Law Enforcement Bulletin, 59*, 6–12.

Walker, S. (1992). *The Police in America*. New York: McGraw-Hill.

Weaver, G. (1992). Law enforcement in a culturally diverse society. *FBI Law Enforcement Bulletin, 61*.

Wiseman, R. L. (1995). *Intercultural Communication Theory*. International Communication Annual, Vol. XIX. Thousand Oaks, CA: Sage Publications.

SPECIALIST
AND DEVELOPMENTAL TRAINING
The Post Hoc Approach

INTRODUCTION

August Vollmer's statement quoted in Chapter 1 exemplifies only a fraction of the problems associated with the need for specialist and developmental training for police officers. To the multitude of tasks and fields of expertise identified by Vollmer, we have to add two additional factors. One is the unprecedented development in technology, the other ideas embedded in the philosophy of community policing. Although developments in technology have had both positive and negative impacts on policing, one of the main tenets of the COP philosophy, the call for decentralization and creation of generalists rather than specialists, deprives policing of the chance to attain the status of a highly reputable profession.

The goal of this chapter is to present some programs offered by various organizations and police departments under the guise of specialized or developmental training. Some of the programs appear to be quite serious in nature, but in most cases, the examples illustrate a less-than-serious approach to specialized training.

SPECIALIZATION

O. W. Wilson studied the relationship between an effective organizational structure of police agencies and specialization. Although he did not find much of benefit in specialized units for smaller police departments, since their patrol officers appear to be jacks of all trades, he identified a number of advantages for large police agencies (Wilson and McLaren, 1972):

- Placement of responsibility
- Development of expertise
- Promotion of group esprit de corps
- Increased efficiency and effectiveness

Specialization is clearly an indicator of status in any profession; indeed, it is rather difficult to think of a reputable profession in which one does not find specialized fields. From medical doctors through school teachers and college professors to lawyers, the initial label is followed by an explanation of a given specialization, so a doctor is not a doctor but a pediatrician or an anesthesiologist, a professor is a professor of criminal justice, and a lawyer is a criminal defense lawyer.

Proponents of the idea of generalist point to a number of problems associated with specialization. It appears to (Swanson et. al., 2001):

- Create increased friction and conflict between units
- Create loyalty to the specialized unit instead of to the department
- Contribute to a decrease in overall job performance due to job factionalism
- Hamper the development of a well-rounded police program.

Although the idea of expertise appears to be more and more crucial in the performance of police tasks in the era of technological advances and new breeds of criminals who transcend not only the borders of neighborhoods and cities but also those of countries and continents, endorsers of the generalist idea seem to hold a very narrow perspective of the dynamics in any work environment. If proponents had conducted preliminary research into other work environments, such as hospitals or universities, they would have found the same frictions, conflicts, overall decrease in job performance, and lack of support for the development of well-rounded programs. Some might actually argue that these more sophisticated establishments generate more conflict and friction than does any highly specialized police agency. However, it is rather difficult to find a voice advocating despecialization in the medical profession or abandonment of their areas of expertise by college professors. The reason for this can probably be traced to basic common sense: The more effort one devotes to studying and learning a given topic, the better insight one gets.

Regardless of the field of work or profession, the meaning of the word *expertise* clearly points to a narrowly defined theme, not to a multitude of skills in a number of fields.

What makes us unique as human beings is our interest in different fields, interests that are further enhanced, developed, and defined through education and training. During the basic academy training, police officers receive a general overview, at best, of most of the topics and issues that are inherent to their profession. In a way this overview can be compared to an introductory course in criminal justice—an overview of the system. No student, after taking the introductory course, can claim or even think about claiming that he or she is now ready to work in the field of criminal justice. It takes another 117 credits (based on the average 120 credits required for a bachelor's degree) to get an undergraduate degree. Even after graduation the student is still barely called a professional. In most professions, an advanced degree is required before attaining the label of practitioner or professional. Furthermore, to be referred to as a psychologist, for example, one must complete a doctoral degree.

Some avid followers and endorsers of the generalist idea might claim that one cannot compare police officers to psychologists. But the philosophy of COP endorses the concepts of problem solving, conflict resolution and crisis management—concepts taken directly from the jargon of other professions, specifically psychology. If one endorses the advanced concepts from another profession, he or she has to adapt the tools that that profession uses to prepare their apprentices. These tools are extensive education and training and with emphasis on specialization that generates the desired expertise. Psychologists are not just psychologists, they study to become clinical psychologists, child psychologists, organizational psychologists, and so on. The status attached to their profession is associated with their field of expertise.

Police officers have and will continue to suffer from a very low occupational status label attached to a profession that should be regarded as highly as any other. Unfortunately, if the generalist approach to training prevails and furthermore, gains momentum in American policing, the suffering will not only continue but will increase even further.

Not only do we not give our officers the tools necessary to perform their jobs, as basic academy training cannot offer them these tools (if it continues to offer training modules inadequate in both length and content), but specialized and developmental training often creates the impression of further deterioration of the idea of professionalism in law enforcement.

The length of the training in itself is seldom a fully inclusive indicator of the quality of a given training module. However, coupled with answers to the following questions they give a clear picture of the quality of the specialized training: (1) what?, (2) when?, (3) where?, (4) who?, and (5) by whom? For example, the Dutch police force recognizes the need for specialized training by offering a lengthy training module for officers who specialize in surveillance which (described in further detail in Chapter 14) provides some productive answers to these questions. The author would add an additional query to the list—"how much?"—as the cost of training programs appears to be one of the main factors contributing to the grim picture.

The multitude of topics and themes covered in specialized and developmental training points to the complexity of the policing profession. This complexity necessitates a serious and structured approach. From highly conceptual skills such as problem solving to highly technical skills such as firearms, the skills required of police officers are endless and diversified. Even a succinct overview of the themes illustrates the intricacy of the police profession.

After a review of some of the programs offered (including the cost), an example of the phase approach to training, introduced by the Greensboro, North Carolina Police Department, will be analyzed as a desired model for replication, followed by some more generic recommendations.

POST HOC TRAINING

One of the more prominent events that highlighted the need for specialized training can be traced to the early 1960s. In August 1966, an incident occurred in Austin, Texas, that, contrary to other incidents, pushed law enforcement toward assessment of their capabilities in handling high-risk situations. After killing his wife and mother, Charles Whitman went to the rooftop of the University of Texas and began a shooting spree, killing 15 people and wounding 30 others. This event contributed to the establishment of special police teams to handle high-risk situations.

As for today, when one analyzes the themes of police specialized training, it appears that new trends such as the increase in violent crimes, high-technology crimes, the increased number of high-risk repeat offenders (an outcome of prison overcrowding), the overall sophistication of the criminal element, diversity-related issues, and a host of additional problems which create the need for a specialized and developmental training approach generate, at best, what the author would call a *post hoc training* approach.

Thibault et al. (1998), in their book *Proactive Police Management*, identified some of the newer topics that have been covered in police training (pp. 329–331):

1. Stress training
2. Dealing with terrorist activity
3. Domestic crisis intervention
4. AIDS awareness
5. Domestic violence training
6. Stalking
7. Tactical operations
8. Diversity
9. Bias crimes
10. Americans with Disabilities Act
11. Computer/video simulation
12. Interactive teleconferencing/cable TV

When describing police subcultures, police scholars refer to the concept of post hoc morality when dealing with explanations for unethical or questionable behaviors. The post hoc morality provides an alibi, explanation, and/or a justification for an officer's behavior after the fact (Crank, 1998). The quality and quantity of various specialized and developmental training courses seem to be providing a similar outlet for police agency as the adoption of post hoc morality. There is, indeed, an element of alibi, explanation, and justification in the various specialized and developmental courses offered to law enforcement officers; there is, however, no trace of an element of true expertise.

A two-day course dealing with diversity or with bias crimes can certainly serve as an alibi for a department that needs to enhance its officers' people skills or prepare for accusations of indifference toward a certain group of victims. However, it will not provide adequate tools to deal with these problems, not even in a semieffective manner.

The author, herself a college professor, educator, and trainer, cannot possibly envision any effective training module on profoundly complex topics such as the ones identified by Thibault et al. conveyed in one-, two-, or three-day segments. After 15 weeks of instruction for 2½ hours a week, a total of about 38 hours (38 to 40 hours of instructions are considered to be the average length of a course in any college environment), both students and professor have a feeling of accomplishment. However, it would be rather presumptuous to claim any type of expertise in a given area of instruction. Ethical behavior, domestic violence, or gang awareness training cannot be delivered effectively during the time allocated for instruction by some of the providers. All they provide is an alibi for a given police organization and a false sense of security for officers in the field.

SPECIALIZED TRAINING

The major drive for police agencies to send its officers for specialized and advanced training appears to be related to law enforcement standards mandated by each state's POST commission. Usually, each commission mandates a number of hours of certified in-service training annually in addition to firearm recertification or requalification. For example, every person employed as a full-time law enforcement officer in Virginia must attend 40 hours of certified in-service training within two years of completion of a basic school. Absences from in-service training prevent recertification with the Department of Criminal Justice Services. Every hour missed must be made up. Compulsory in-service standards include the following training modules:

- Mandatory: legal training (4 hours)
- Career development (16 hours)
- Elective (20 hours; no more than 8 hours of firearms training are permitted as an elective subject)

Firearms recertification standards require that every officer who carries a firearm in performance of his or her duty must qualify annually, attaining a minimum qualifying score of 70 percent. This qualification must occur on one of the following courses (North Virginia Criminal Justice Academy, 1999):

- The Virginia modified double-action course for revolvers
- The Virginia modified double-action course for semiautomatic pistols
- The Virginia 50-round tactical qualification course for revolvers and semiautomatic pistols
- A special weapons course using shotguns and/or similar weapons

What is interesting and notable from a review of the recertification standards is the fact that an officer has to attain a minimum score of only 70 percent, and then of course there is a public outcry when things happen under stress in real-life situations. Unfortunately, most academies require only this minimum score of 70 percent. The question that must be addressed is whether or not the general public is fully aware of the fact that because of a lack of adequate resources, we deposit the most powerful tool in the hands of our police officers but do not allocate enough resources to enable him or her to qualify with a minimum score of 90 percent.

The specialized and advanced course offered to and by police agencies can be divided into a number of categories:

- Offered by in-house municipal police academies
- Offered by federal/state/regional law enforcement agencies
- Offered by private individuals
- Offered by private organizations

The training is offered externally either on the premises of the offering agency or in a designated location, some of which is customized to the needs of the clients and brought directly to the department, either through distance learning techniques, television networks, or in person. Some of the training modules are offered exclusively to public law enforcement officers, some are open to both private and public law enforcement, yet others are open to the general public. This mix of publics in itself appears to contribute in a significant manner to the problems associated with the effectiveness of any given instructional module.

Note: In the discussion of specific training programs that follows, the cost of a given module is provided as an illustration of how much money an agency would have to spend to provide a given training to officers. The costs change constantly; however, some estimates can be made based on the costs quoted in this chapter.

R.E.B. Training International

R.E.B. Training International specializes in training and consulting on the subject of managing aggressive behavior. Since being established in 1983 it has offered programs and training modules to some 4000 agencies. These agencies and organizations include police enforcement agencies, security, academic, health care, military, and federal agencies. The stated mission of this training provider is

"to teach individuals how to protect themselves from injury and control other individuals without causing them harm. To resolve conflicts decisively and diplomatically regardless of age, size and strength."
Courses offered are (*http://www.rebtraining.com*):

- *MOAB:* A CD-ROM-based course that focuses on providing employees with the skills to recognize, reduce, and manage upset, verbally and physically aggressive people. In addition, the course minimizes the potential for injury to employees and the people they confront. The courses lasts from one to four days (in-house instructor course), and the CD costs $695 for the first and each additional CD costs $275.
- *OCAT* (Oleoresin Capsicum Aerosol Training): A one-day course in how to use pepper spray (oleoresin capsicum). The price ranges from $195 to $240.

National Institute of Ethics

The National Institute of Ethics defines itself as "the nation's largest provider of training that helps prevent employee misconduct and enhances integrity." With customized programs this organization focuses on training through computer software, audiocassettes, video training tapes, manuals, books, cartoon action characters, and seminars. Its mission is "to enhance professionalism through ethical standards and integrity training." The organization offers the following courses (*http://www.ethicsinstitute.com*):

- *Integrity Leadership:* two-day course
- *Advanced FTO:* three-day course
- *Ethics Instructor Certification:* two- to four-day course

Executive Protection Institute

The Executive Protection Institute is a Virginia-based company that offers courses in protection. The agency trains anyone who is "capable of performance in business and social environments domestically or internationally." The company focuses on protection of recognized world leaders and corporate executives.
Courses offered are (*http://www.personalprotection.com*):

- *Executive/VIP protection:* trains people how to plan preventive strategies in personal protection; guidelines, movements, and checklists. The course lasts two days and costs $395.
- *Managing security systems:* focuses on how to approach the planning, design, and management of a facility's protection systems. The course looks at access control, card encoding techniques, locking systems, alarm systems, sensors, and intrusion detection: basically, anything that has to do with protective systems. It is a two-day course and the price is $395.

- *The Protectors:* handgun defense program for the public or private personal protector. The program content includes shooting styles and techniques, reaction to attacks, multiple targets, equipment, night-vision shooting, and evacuation. The program costs $995, which includes meals, refreshments, snacks, pens, and notepads, and the course lasts three days.

Street Survival Seminar

The Street Survival Seminar was founded 24 years ago and is a three-day course. On days 1 and 2 the classes focus on how to perform more effectively in high-risk situations, detection of disguised and concealed weapons, use of force, street-gang tactics, tactics to protect one's self and family, real-life confrontations, and unmasking drug couriers. During the third day, participants learn in court tactics, family communication, report writing, recovering from damage, civil lawsuits, traumatic events, and destructive behavior managing skills (*http://www.calibrepress.com*).

Law Enforcement Television Network

The Law Enforcement Television Network (LETN) offers multimedia to meet the needs of today's law enforcement agencies. The STTAR (Specialized Training Testing and Record) system combines management and communications with multimedia training. It is designed to enhance a training program in two important areas: first, to provide a self-paced program in which testing can be done anywhere, and second, to provide a more effective self-managed training program.

The program functions by allowing a person authorized use of the system. When an ID number has been given, the person can monitor and control his or her training activity. Courses are shown via satellite or as videotapes and can be seen anywhere in the world. Tests can also be taken all over the world. The tests are updated monthly via modem and are graded automatically and stored in a confidential file. The records are stored in the student's personal file.

Courses are offered in several different categories (*http://www.letn.com*):

- Corrections
- Courthouse
- Dispatch
- Investigations
- Investigative techniques and drugs
- Leadership
- Legal issues
- Management
- Patrol
- Personal development

- Personnel
- Public education
- Safety
- Supervision
- Supervisors
- Tactical issues
- Training

From these broad categories, people can choose from almost any law enforcement topic. The courses can last from a few minutes to several hours. Based on the length of some of the courses, especially the shorter ones, they are more appropriate for brief roll-call training, as a refresher rather than a new concept medium.

Hutchinson Law Enforcement Training

The Hutchinson Law Enforcement Training (HLET) program started its training in 1993 and has trained officers from over 900 federal, state, county, local, and university police agencies. The program offers training in subjects that are usually not introduced during basic academy or/and in-service training.

Several HLET seminars are offered (*http://patriotweb.com/hlet/*):

- Background investigations
- Body language and interviewing techniques
- Community policing
- Confidential informant operations
- Criminal intelligence functions
- Drug interdiction
- Interview and interrogation methods
- Investigative techniques
- Narcotic investigations
- Police ethics and diversity training
- Raid planning, preparation, and execution

All of the seminars last two days, with approximately 16 hours of instruction.

Tactical Response Training

Tactical Response Training specializes in advanced developmental and tactical training for law enforcement agencies. The courses offered are designed for SWAT or special operations personnel. The courses are intense and begin with static instruction, then progress to fluid application, and end with dynamic drills.

The program has trained personnel from over 50 agencies, including European. Most courses costs between $75 and $100 a day.

Courses offered include (*http://www.tacticaltraining.net*):

- *MP5/SMG Operator:* a three-day course in which the focus is on operating an MP5 or Colt SMG. Also covered are nomenclature, cover and concealment, marksmanship fundamentals, dim-light shooting, practical handling techniques, SMG weapon retention, care and cleaning, and range work.
- *Dignitary Protection Operations:* three days long. Focuses on tactics and how to carry out dignitary protective details. Topics covered include causes of assassination, methods of attack, protective details, formation movements, motorcades, and practical exercises.
- *Vehicle Assaults and Hostage Rescue:* a one-day course in how to understand procedures, equipment, and personnel needs to effectively carry out vehicle assaults. Topics include tactics, rescue techniques, and live-fire drills.

North American SWAT Training Association

The North American SWAT Training Association (NASTA) offers a 40-hour training program that can be tailored to individual personal needs (*http://www.nasta1.com*). The classes are usually presented in the basic SWAT school and include:

- Physical fitness assessment
- Firearms assessment and training
- Scouting and mission preparation
- Slow-search class and practicals
- Tactical firearms
- Dynamic entry class and practicals
- Physical exertion firearms
- Responders to critical incidents
- Obstacle course

NASTA also offers an advanced SWAT school program. This is designed for the experienced tactical officer. Classes offered in this 40-hour programs include:

- Team-building exercise
- Scouting and mission preparation
- Night-movement techniques and practicals
- Slow-search class and practicals
- Advanced search techniques
- Breathing tools/distraction devices

- Dynamic entry class and practicals
- Officer-down techniques and practicals
- Hostage/barricade class and practicals

In addition, NASTA has developed a third program called QUAD (quick action deployment) training. This program focuses on how to respond immediately to a crisis situation without disregarding an officer's safety. There are two QUAD training seminars, both lasting 8 hours to over one day.

The first, the QUAD training seminar, includes the following themes:

- Preparatory planning
- Incident command post
- Tactical command post
- Evacuation locations and operation
- Media and parents
- Debriefings
- After-action reports
- Poststress issues
- Quad training presentation and video

The second course, QUAD training with practical exercises, includes the following discussions:

- Preparatory planning
- Evacuation locations and operation
- Quad training presentation and video
- Quad training practical exercises

International Association of Chiefs of Police

The International Association of Chiefs of Police (IACP) organization offers a vast array of courses nationwide. This professional organization should serve as a model for police organizations involved in developmental and specialized courses. The courses offered include topics and themes that are relevant, up-to-date, and of primary importance. The instructors are highly qualified, but both the cost involved and the length of the modules make them more appropriate for police departments willing to adopt the phase approach to training (that is discussed and analyzed later in the chapter).

Following is a sample of seminars currently offered (*http://www.theiacp.org/training*):

- *Ethical Standards in Police Service*. This two-day course provides law enforcement officers with the tools needed to maintain and improve ethical standards. Participants are taught how to address certain specific ethical issues that confront particular levels of a police agency, ethical standards and success, and the limitations of external controls on police behavior. The price is $285 for members and $385 for nonmembers.

- *Advanced Crime Analysis.* This three-day course is intended for officers with experience in the area of crime analysis. The course includes computer exercises and analysis, and discussion covers crime bulletins, analytical reports, mapping, and GIS-based analysis. The price is $360 for members and $460 for nonmembers.

- *Reducing School Violence.* This course focuses on the communication between law enforcement and school administrators, and areas covered include psychological and behavioral foundations, cultural and social dimensions, family and community life patterns, public action and legal aspects, aggressive students, school search and seizure, and community policing initiatives in the school community. The course lasts two days; the price is $150 for members and $250 for nonmembers.

- *SWAT I: Basic Tactical Operations and High Risk.* This course focuses on SWAT responsibilities, operations, concepts, equipment, gas teams, scouting procedures, and operational procedures. The course is open to sworn officers or full-time employees of law enforcement agencies, and the participants learn how to handle barricaded suspects, design special arrest teams operations and concepts, and employ basic rappelling techniques. The duration is five days; members pay $495 and nonmembers pay $595.

- *Value-Centered Leadership: A Workshop on Ethics.* This course lasts two days and costs $285 for members and $385 for nonmembers. The course focuses on the dynamics of ethics and the values within a law enforcement agency, and on communication and interaction with the surrounding community.

- *Investigation of Incidents of Excessive/Deadly Force.* This course is intended for chiefs, commanders, supervisors, and other people who may handle deadly force allegations. It covers how situations in which deadly force is an issue should be handled in order to prevent negative outcomes. The course lasts four days; members pay $425 and nonmembers pay $525.

- *SWAT II: Advanced Tactical and Hostage Rescue Operations.* This course is designed for SWAT officers who have prior training. Includes hostage rescue techniques, coordinated multiple target selection and neutralization tactics, multiphase special operations, weapons handling, advanced training methodology, and complex command and control concepts. In addition, participants learn the modern techniques of pistol, submachine gun, and combat shooting. The course lasts five days; the price is $495 for members and $595 for nonmembers.

- *SWAT Supervisors' Tactics and Management.* This course is designed for SWAT commanders, supervisors, and command personnel. Areas covered include selecting personnel, civil liability issues, maintaining personnel, selecting appropriate weapons, and testing and training SWAT units. The duration is four days; the price is $400 for members and $500 for nonmembers.

- *Investigation of Computer Crime.* This course focuses on computer crime and covers areas such as theft of information, credit card fraud, computer hacking, intrusion, theft of intellectual property, and child pornography. Participants will attain knowledge on cyber terrorism, legal issues, and Internet security. The course is given over three days; member price is $360 and the nonmember fee is $460.

- *Risk Management for Law Enforcement Agencies.* This is a two-day course that focuses on an agency's risk needs, risk management structure, civil litigation, risk management information systems, and loss prevention and recovery programs. The price is $285 for members and $385 for nonmembers.

GIS/GPS Training

One of the most popular themes in specialized and developmental training is the GIS/GPS crime-mapping technology. Computer mapping is a powerful tool, merging raw street intelligence and advanced data tables into a medium that can be utilized by both large and small agencies.

Three publicly funded agencies currently offer free training courses to help law enforcement personnel in crime mapping technology.

1. *Crime Mapping and Analysis Program:* in Denver, Colorado; a two-week introductory course and several advanced courses

2. *Carolinas Institute for Community Policing:* in six cities throughout South and North Carolina; a series of courses that emphasizes the importance of crime mapping technology in policing

3. *Crime Mapping Research Center:* in collaboration with the Office of Community Oriented Policing; training in crime mapping at regional community policing institutes throughout the nation

The course listing includes (*http://www.ojp.usdoj.gov/cmrc*):

- *Crime Mapping for Community Policing and Problem Solving* (4 hours, geared toward officers, community, others interested in the basics of crime mapping)

- *Mapping for Managers* (4 hours, geared toward administrators and managers who want to know about crime mapping, what to ask for, and what to expect)

- *What Is Crime Mapping?* (8 hours, geared toward analysts, officers, community, others who want a little more in-depth look at crime mapping)
- *Integrating GIS into an Organization* (8 hours, geared toward analysts, officers, others who play a role in implementing crime mapping in their agency)

The Crime Mapping and Analysis Program

The Crime Mapping and Analysis Program provides technical assistance and training to law enforcement agencies. The training project begins with a one-week session that introduces participants to crime mapping. The next session is designed to focus on the packages generally used for standard crime mapping. The final part is an applied project in analyzing crime in which participants can try their skills.

North Crest Resource Management

North Crest Resource Management (NCRM) is an organization that has specialized in land management consulting services since 1982. They offer seminars in California on GIS/GPS developments in the world. Most of them last only one day (*http://www.ncrm.com*).

GPS/GIS Training

GPS/GIS Training offers the following seminars (*http://www.gps-training.com*):

- *Training Seminar for GPS/GIS Training.* This is a two-day course, costing $285. The course focuses on mission planning, job design, feature list creation, static/dynamic techniques for collecting points, lines, and areas, PC file transfer, coordinate system transformation, GIS mapping/exporting, and plot output.
- *Training Seminar for Single- and Dual-Frequency CPS.* This course lasts five days and costs $595. This course is an extension of the course above and gives hands-on experience to participants.
- *Training Seminar for PLGR and MC5 Mapper GPS.* This is a two-day course that costs $285. This is a course that is designed for the use of either a PLGR or an MC5 mapper as a GPS/GIS mapping tool.

Institute for Transportation Research and Education

The Institute for Transportation Research and Education (ITRE) offers training in GIS/GPS mapping. The courses are designed to focus on the advances in the technology and thereby give the participants relevant knowledge to prepare them for future work with electronic mapping. The program was founded in 1989 and offers training to state, local, and private organizations. Several different seminars are offered, lasting from two to five days. The price varies from $399 to $999 (*http://www.itre.ncsu.edu*).

Mississippi State University

Mississippi State University (MSU) occasionally invites governmental agencies to participate in free one-day workshops on GIS/GPS training (*http://www.ur.msstate.edu*).

Montana Tech (University of Montana)

Montana Tech offers a four-semester program that focuses on GIS/GPS training. The program is called "Technology Program Geographic Information System/Global Positioning System." Its aim is to enable students to use the GIS software to process and produce maps and to prepare the student for jobs in this technological area to assist various law enforcement agencies.

This intensive program is composed of the following four semesters (*http://www.mtech.edu*):

- *First semester:* CAD I, Introduction to Writing, Principles of Speaking, Microcomputer Software, Intermediate Algebra, and General Psychology
- *Second semester:* Introduction to Surveying, English Composition, GIS Software Applications, Data Acquisition for GIS/GPS, Data Files Management I, Employment Strategies, and Geometry and Trigonometry
- *Third semester:* Plane Surveying, Watershed Mapping, Global Positioning System, and GIS Software II
- *Fourth semester:* Map Design, Introduction to Statistics, GIS/GPS Practicum, Computer Repair and Maintenance, and Introduction to Programming

Other Resources

There are numerous additional organizations that offer training in the area of GIS/GPS as the field has become one of the "hottest" topics for law enforcement. Unfortunately, the degree of expertise that those organizations provide is not comparable. Police agencies should be extremely selective when choosing a provider in this futuristic field. Some additional providers of crime mapping training include CMT–Corvallis Microtechnology Inc., CompassCom Inc., ESRI Education Solutions, System Dividends, and Hamilton Geodetic Services.

Any person or agency interested in police training can use the advances in technology, specifically the Internet, to identify a myriad of courses offered to law enforcement professionals. One of the best sites is *http://www.officer.com*, which offers updates on training courses given in various areas of the country. The sample of courses reviewed above is by no means an inclusive representation of training opportunities. However, it certainly is representative of the topics, lengths, and prices of the themes and trends.

Based on the overview above, the overall impression can be summarized as follows:

1. Providers of specialized and developmental training (both public and private) have full awareness of the critical topics, themes, and skills that need to be developed, enhanced, or introduced.

2. The same providers exhibit a lack of awareness regarding some critical issues:
   - *Necessary length of a given module.* Most modules are offered in what can be called an abbreviated version, less than what is required to deliver effective training.
   - *Effectiveness of a given module.* Delivered in its abbreviated form, a module is not only limited in its effectiveness but may prove disastrous to officers in the field, who are under the impression that they actually received adequate training.
   - *Charge of negligent training.* Such a charge can be brought against the department that exposed its officers to the abbreviated training module.
   - *Mix audiences.* Offering the same training modules to audiences representing both private and public sectors detracts from the overall effectiveness of the module and sometimes invalidates the concepts being taught.

3. Some providers appear to be motivated by the cost of the module rather than by its effectiveness and neglect to foresee the potential disastrous outcomes of inadequate training.

GREENSBORO, NORTH CAROLINA, PHASE APPROACH TO TRAINING

Wood (1980) presented an interesting approach to specialized training developed by the Greensboro, North Carolina, Police Department. This approach should be taken into consideration as a model for baseline training for all specialized and developmental courses. Adopting Greensboro's method would in turn provide some extremely fertile ground for the courses, offered by highly qualified providers. The department recognized the need for a multiphased approach to training and development of police officers and established a standard training program for each of the following phases.

Phase I: Recruit Training

Phase I is composed of two subphases:

1. 560 hours of basic academy training. Upon its completion, the officers receive a sworn status of Police Officer I.

2. Field training under the supervision of a police squad leader. Upon successful completion, the officers receive patrol duty assignments.

Phase II: Intermediate Training

Phase II lasts approximately one and a half years and consists of:

1. On-the-job training with the assigned field division
2. Quarterly in-service training composed of 8-hour training modules, consistent with the needs of the department
3. Monthly 2-hour modules presented during roll-call sessions

Phase III: Advanced Training

Phase III marks the promotion period for Police Officer I, when upon completion of two years of service, an officer is eligible for promotion to Officer II. The promotion is based on the outcomes of a written competency test based on the knowledge gained by the officer during phases I and II. Phase III is composed of the following training components:

1. Quarterly in-service training (as in Phase II)
2. Monthly roll-call training (as in Phase II)
3. Advanced training through periodic seminars and workshops having developmental and performance oriented frameworks.

To afford each officer equal-opportunity training, assignments are made on the basis of seniority and are rotated, enabling each officer to attend one training module before a senior officer is assigned to a second module.

Phase IV: Developmental Training

Phase IV essentially enables Police Officer II to participate in two developmental programs:

1. *Career developmental stages*
   - A short-term (six- to eight-month) temporary assignment to one of the investigative or staff divisions
   - 30 days of orientation (during the temporary assignment) within the area selected
   - Supervised on-the-job training
   - Attendance in seminars and workshops related to the specialty area
2. *Police squad leader program.* Officers go through an established selection procedure and are promoted to police-squad-leader status. In this position an officer serves periodically as training coach for newly assigned officers and may also act as the squad supervisor in the absence of regularly assigned supervisor. The officers receive training in the following areas:
   - Coach training
   - Instructor training
   - Instruction to supervision

Phase V: Specialist Training

Phase V is designed for officers assigned to specialized units such as criminal investigation, vice/narcotics, youth, traffic, and so on, and officers who participate as members of special teams such as SWAT, undercover recovery, or bomb squad. This phase offers the following training opportunities:

1. Seminars and workshops, offered by external agencies such as the Northwestern University Traffic Institute, Southern Police Institute, FBI National Academy, and others
2. Quarterly in-service training sessions

Phase VI: Supervisory Training

Phase VI is designated for officers promoted to the rank of sergeant. This phase is composed of the following stages:

1. *Practical supervisor's course.* This is offered by two primary sources: (1) the Institute of Government in Chapel Hill, North Carolina, and (2) the North Carolina Governor's Highway Safety Program, both offering a management-administration course of approximately 200 hours of instruction.
2. *In-house supervisory orientation program.* This is designed to allow supervisors to familiarize themselves with departmental policies, programs and procedures.

Thibault et al. (1998) suggested an addition to this supervisory phase VI by supplementing training with the "Model of First-Line Manager (Supervisor)," adapted by them from the revised supervisor's course curriculum being used in New York State by the Division for Local Police, Municipal Police Training Council, in Albany, New York.

The following model covers themes and topics related to first-line supervisory skills. Assignments are linked to daily class topics, and the results are graded on a pass/fail basis, while all "fails" have to be repeated until they become "passes."

1. Registration and overview
2. Role of supervisor and leadership skills
3. Wellness program
4. Supervisor communications and media relations
5. Civil liability
6. Evaluation, counseling, and discipline
7. Intermediate law
8. Crime prevention and community relations
9. Stress management
10. Personnel issues
11. Use of deadly physical force
12. Report review
13. Supervisor as trainer

The program is designed as an intensive residential program, where the students prepare for each class with out-of-class readings and homework assignments. The classes are offered both day and night. It is recommended for officers prior to assuming the duties of supervisors, as part of a one-year civil service probationary period before certification as a supervisor (Thibault et al., 1998, pp. 335–337).

Phase VII: Management Training

Phase VII is linked to an officer's promotion to the rank of lieutenant. This phase consists of the following stages:

1. Initial training consists of either a three-month police administration course at the University of Louisville Southern Police Institute or a three-month traffic police administration training program offered by the Northwestern University Traffic Institute.
2. From the rank of captain and above, officers are eligible to apply to the FBI National Academy.
3. In addition to these major programs, lieutenants, captains, and majors may participate in a variety of seminars, workshops, and conferences.

The phase approach to training is a flexible approach. Specific programs in any phase can be changed and customized for the departmental needs. Using these concepts provides a number of benefits to the department (Thibault, 1998, p. 339):

1. It helps ensure that officers have the skills and knowledge to perform at each level of their development in the department.
2. Officers continue to be engaged in career development on a timely basis
3. Officers know what they can expect from the department in the area of training and development, which can contribute greatly to the officers' morale.
4. A fair and equitable system is established in which every officer receives approximately the same level of training.
5. The phase approach helps in the coordination and planning of the department's overall training program and helps to avoid pitfalls in administering the program.

In addition, this author has identified a number of additional attributes that the phase approach could possibly offer as value-added contributions to a department.

1. It provides an excellent source for defense from potential liability suits as it not only shows a well-planned, researched, and updated approach to training but also a true antidote to post hoc training concepts.
2. It can save money by allowing identification of the best providers, contacting them with the long-term plan for the entire department and negotiating a better price package based on the long term and the volume of the commitment.

3. It allows for a certain uniformity of police officer training, a factor that contributes not only to development of an expertise but also to the overall view of policing as a true profession. After all, the basic curriculum of medical or law schools around the country does not differ that much in its content and length. A phase approach to specialized training provides for a similar and progressive baseline.

NORTHERN VIRGINIA CRIMINAL JUSTICE ACADEMY IN-SERVICE TRAINING

In addition to the agencies that offer in-service specialized and developmental training mentioned in the phase approach, one agency, the Northern Virginia Criminal Justice Academy, deserves special attention for the quality of its in-service training. Established in 1965, in its 35 years of service it has grown from the original founding members of Arlington County, Fairfax County, and the city of Alexandria to include 17 different agencies. Presently, over 100 programs are offered to approximately 2000 criminal justice professionals.

Topics of instruction have been selected carefully to meet the needs of the participating agencies and reflect the current societal changes facing law enforcement today. The emphasis is not only on keeping up with topical trends but keeping ahead of them. The academy exemplifies a proactive approach to specialized training and is fully capable of providing a support system for the phase approach. The courses offered are grouped by major areas and include specialized and developmental courses in the following areas:

1. Patrol
   - Advanced Patrol Officers School
   - Decision Making for Law Enforcement
   - Field Training Officer Seminar
   - Law Enforcement In-Service
   - Mental Preparedness for Law Enforcement
   - Narcotics Investigations and Interdiction
   - Robbery Pickpocket Seminar
   - Vehicle Stops
2. Investigations
   - Blood Stain Patterns
   - Child Abuse Investigations
   - Electronic Surveillance Techniques
   - Evidence Collection (Basic)
   - Homicide Investigations (Basic)

- Homicide Investigations (Major Case Management)
- Interviews (Basic)
- Photography (Advanced)

3. Traffic
   - Accident Investigation (Basic)
   - Accident Investigation (Advanced)

4. Legal Considerations
   - Legal Day: Law Enforcement
   - Legal Day: Corrections

5. Skills
   - Bicycle Operators
   - Computer Operations (Basic)
   - Defensive Tactics Instructor
   - Defensive Tactics Instructor (Recertification)
   - Firearms Instructor (Basic)
   - Firearms Instructor (Recertification)
   - Ground Fighting and Defense
   - Instructor Developmental (Basic)
   - Instructor Developmental (Advanced)
   - Instructor Developmental (Recertification)
   - Law Enforcement and the Internet
   - Power Point: Level 1
   - Radar Operator (Basic)
   - Radar Operator (Recertification)
   - Radar Instructor (Recertification)
   - Standardized Field Sobriety Testing
   - SWAT School Basic
   - VCIN/NCIC
   - VCIN/NCIC (Recertification)

6. Driver Training
   - Driver Training Instructor (Recertification)
   - Driver Training Refresher
   - Motorcycle Operations (Introductory)
   - Motorcycle Operations (Basic)
   - Motorcycle Operations (Refresher)

7. Leadership, Management, Supervision
   - Career Development Seminar
   - Challenges in a Changing Workplace
   - Crisis Management I

- First-Line Supervisors (Basic)
- FTO Issues: Command Overview
- FTO Supervisors School
- Hostage Negotiations
- Leadership Series: Challenge of Transition
- Leadership Series: Advanced Techniques and Methodologies
- Leadership Series: Command Issues
- Pursuing Excellence in Law Enforcement
- Project Leadership Retirement Issues for Law Enforcement

8. Specialized Topics
   - Survival Spanish
   - Survival Sign Language
   - Verbal Judo

9. Support Functions
   - Telecommunications (Basic)

The Northern Virginia Criminal Justice Academy is not the only training academy that offers some excellent in-service training modules. There are a number of academies around the country, especially state and regional, that provide some extensive developmental training as well. However, the examples provided above illustrate a wide range of topics that can definitely be taken into consideration by other academies that do not offer many extensive modules.

The length of a course ranges from 8 to 80 instructional hours (Northern Virginia Criminal Justice Academy, 1999). There appear to be two disadvantages in the academy's approach to training. One, discussed in Chapter 6, relates to the qualifications of instructors. One of the required qualifications of the instructors is a certain level of physical fitness. Again, the reader is reminded that certain conceptual topics need not be conveyed through arm muscles. The second drawback relates to some of the themes missing from the curriculum, especially in the area of specialized topics such as high-technology or white-collar crime, but also more operational training in the area of civil disobedience, dealing with special populations (mentally ill), and crowd control. However, the academy's reference manual, compiled to evaluate law enforcement performance outcomes as they related to mastery of the topics covered during the instructional modules, should serve as a blueprint for every training academy and certainly shifts attention from these two drawbacks. Furthermore, the curriculum is pretty impressive and should be emulated by other academies.

CONCLUSIONS

What constitutes an interesting phenomenon is the fact that many in-house academies that do offer basic academy and even control the FTO training tend to stay away from offering advanced and specialized training. The author has discussed this issue with a training director of one of the academies, and inquired why the academy does not provide specialized, developmental, and advanced training to its employees. The answer received was one of hesitation; it appears that traditionally, officers were sent to other providers for external specialized training without a good and valid reason. Academies, especially in-house academies, could save money and boost morale by providing specialized training modules internally using in-house as well as external experts. This theme is addressed further Chapter 15.

A final note: While researching existing materials for this chapter, the author came across a book published in 1974, almost three decades ago. In a book entitled *The Police and the Behavioral Sciences*, Steinberg and McEvoy describe five innovative training programs instituted by five different police departments.

1. In-Service Child and Juvenile Training Program for Patrol Officers, developed by the Richmond, California, Police Department
2. UCLA Community–Police Relations Training Program
3. Oakland Model for Dealing with Family Crisis: Specialists Model
4. Conflict Management Model, developed in Dayton, Ohio
5. Cincinnati Human Relations Training Program

A number of very interesting concepts for consideration were identified by Steinberg and McEvoy almost 30 years ago: first, the need for specialists; second, the need for extensive in-depth training in topics of a sensitive nature (e.g., they recommended that the human relations program be included in the basic academy training with two 40-hour modules); and third, the impact on learning of a physical training environment. The latter concept is discussed further in Chapter 15.

SUMMARY

The importance of specialized training for police organizations cannot be overemphasized. The contribution of this chapter to the list of critical issues in police training cannot be overstated either. Arguments for and against specialized training were presented and discussed from both a purely theoretical angle and then by looking at a number of specialized training modules offered by various agencies and private entities. While some of the programs presented and the amount of detail included might appear unnecessary at first glance, the need to understand the seriousness of the problem necessitated such a graphic approach.

What appears to be striking about Steinberg and McEvoy's findings is the fact that for decades progressive ideas regarding police training are being publicized, but most police training academies and police agencies do not seem to be bothered by the obvious, and instead of striving toward professionalism, they seem to endorse with open arms one-, two-, or three-day post hoc training (i.e., "alibi") courses, and continue to wonder why every police officer knows exactly how many years, months, and days he or she has left until retirement.

REFERENCES

Anonymous. Retrieved September 20, 2000 from the World Wide Web: *http://www.officer.com.*

Crank, J. P. (1998). *Understanding Police Culture.* Cincinnati, OH: Anderson Publishing.

Executive Protection Institute. Retrieved September 20, 2000 from the World Wide Web: *http://www.personalprotection.com.*

GPS/GIS Training. Retrieved Sept. 20, 2000 from the World Wide Web: *http://www.gps-training.com.*

Hutchinson Law Enforcement Training, LLC. Retrieved September 20, 2000 from the World Wide Web: *http://patriotweb.com.*

Institute for Transportation Research and Education. Retrieved September 20, 2000 from the World Wide Web: *http://www.itre.ncsu.edu.*

International Association of Chiefs of Police. Risk management for law enforcement agencies. Retrieved September 20, 2000 from the World Wide Web: *http://www.theiacp.org/training.*

Law Enforcement Television Network. Retrieved September 20, 2000 from the World Wide Web: *http://www.letn.com.*

Mississippi State University. Retrieved September 20, 2000 from the World Wide Web: *http://www.ur.msstate.edu.*

Montana Tech. Retrieved September 20, 2000 from the World Wide Web: *http://www.mtech.edu*

National Institute of Ethics. Retrieved September 20, 2000 from the World Wide Web: *http://www.ethicsinstitute.com.*

North American SWAT Training Association. Retrieved September 20, 2000 from the World Wide Web: *http://nasta1.com.*

North Crest Resource Management. Retrieved September 20, 2000 from the World Wide Web: *http://wwwncrm.com.*

Northern Virginia Criminal Justice Academy, 1999. Training Catalogue.

R.E.B. Training International, Inc. Retrieved September 20, 2000 from the World Wide Web: *http://www.rebtraining.com.*

Steinberg, J. L., and McEvoy, D. W. (1974). *The Police and Behavioral Sciences.* Springfield, IL: Charles C Thomas.

Street Survival Seminar. Retrieved September 20, 2000 from the World Wide Web: *http://www.calibrepress.com.*

Swanson, C. R., Territo, L., and Taylor, R. W. (2001). *Police Administration: Structures, Processes, and Behavior,* 5th ed. Upper Saddle River, NJ: Prentice Hall.

Tactical Response Training, Inc. Retrieved September 20, 2000 from the World Wide Web: *http://www.tactical training.net.*

Thibault, E. A., Lynch, L. M., and McBride, R. B. (1998). *Proactive Police Management,* 4th ed. Upper Saddle River, NJ: Prentice Hall.

U.S. Department of Justice. Retrieved September 20, 2000 from the World Wide Web: *http://www.ojp.usdoj.gov/cmrc.*

Wilson, O. W., and McLaren, R. C. (1972). *Police Administration.* New York: McGraw-Hill.

Wood, D. E. (1980). Phase approach to training. *FBI Law Enforcement Bulletin, 49,* no. 5, May, pp. 13–15.

SUPERVISORY AND MANAGEMENT TRAINING

INTRODUCTION

Advanced police training is not a luxury but a necessity in the United States today. The officers who are trained in advanced police courses are a special group of people chosen to handle administrative and tactical/operational situations beyond the scope of basic police activity. Uniqueness, complexity, unpredictability, and the possible consequences of each situation necessitate advanced training.

The goal of this chapter is twofold: first, to highlight some of the problems inherent in supervisory training, and second, to present some of the programs that should be modeled, replicated, and incorporated into various in-house academies. Numerous organizations—law enforcement, academic institutions, and other private entities—offer advanced training for police supervisors. A number of those organizations, such as the FBI National Academy, Federal Law Enforcement Training Center (FLETC), and the International Association of Chiefs of Police, offer a valuable contribution to advanced training by combining a practical law enforcement approach with an academic "touch." However, access to these training opportunities is relatively limited, for a number of reasons, including cost, personnel (who cannot be spared to attend the national academy, which requires a long absence from the force), political interference, ignorance about existing opportunities, and lack of mandatory requirements.

"OVERNIGHT" SUPERVISOR

In many agencies around the country, all it takes to become an "overnight" supervisor is word from the chief. One does not have to qualify by passing a test or going through advanced training. Most readers who have been employed by a state or local agency can offer numerous examples of such overnight promotions, which frequently led to less than desirable consequences for the people involved and the organization itself.

A bigger problem, however, lies in what the author would call a dilettante approach to advanced training offered by some other institutions, some of them of high quality but removed from the reality of life in police agencies, others having no or barely minimal qualifications to be engaged in this business.

Finally, the biggest problem overall appears to be the fact that there is no centralized control or assessment of advanced police training, not even at a minimum level. The POST committees in each state at least provide some minimum standards for basic academy training; unfortunately, most of them do not do it for advanced/supervisory training, which is certainly not less crucial than basic training.

Historically, a number of attempts have been made to mandate or at least recommend some changes in the advanced/supervisory training approach. In 1973, the National Advisory Commission on Criminal Justice Standards and Goals published a volume of recommendations entitled "Police." " Training before promotion" as a policy in any given department was advocated, as well as management programs for middle management. The commission advocated that programs for middle management be integrated with college and business programs (National Advisory Commission, 1973). Operation Bootstrap and the SMIP, both reviewed below, provide an example of the implementation of this recommendation.

Another development took place in the 1970s with an enhanced version of Project STAR (System and Training Analysis of Requirements for Criminal Justice Participants), which was originally developed by the California POST commission. In the enhanced version, four states became involved in the project: California, Michigan, New Jersey, and Texas. Project STAR relied on a systems approach to training and emphasized management by objectives while breaking down police behaviors into specific roles and tasks. Thirteen roles were identified that were supposed to be filled by 33 tasks. The problems with both roles and tasks were summarized by Thibault et al. (1998):

- Much was written in an incomprehensible language using undefined terms.
- Too much emphasis was placed on the latest thinking on management and human behavior.
- Too little emphasis was placed on the practical training on how to actually perform the job.
- Excellent concepts were widely discussed but never implemented.

Professional organizations such as ASLET (American Society for Law Enforcement Trainers) or IDEALIST have attempted for years to provide some guidelines and standards for this cause. Unfortunately, it is rather rare to find many members of these organizations, even among people engaged actively in law enforcement training. It is difficult, if not impossible, to spread the word if there is no serious vehicle to endorse the ideas being presented and developed. Only an organization similar in its nature to POST, which would clearly define and mandate supervisory and advanced training for every state, could make a real difference. Some states do benefit from their POST commission involvement in advanced training; for example, the California POST is heavily involved in addressing the advanced training programs and offers a POST management certificate to graduates of the Sherman Block Supervisory Leadership Institute and the Command College. Unfortunately, examples of such involvement are relatively few. In this chapter we offer a review of a number of advanced training programs, modules and courses offered by various organizations, followed by an analysis of what is lacking and what should and can be accomplished in the field of supervisory training.

ADVANCED TRAINING PROGRAMS

Note: In the discussion of specific training programs that follows, costs are provided as an illustration of how much money an agency would have to spend to provide a given training to officers. The costs change constantly; however, some estimates can be made based on the costs quoted in this chapter.

FBI National Academy

The National Academy, located in Quantico, Virginia, sponsored by the Federal Bureau of Investigation, offers advanced-level courses for police executives. The goal of the National Academy is to enhance the ability of experienced law enforcement personnel to fulfill their responsibilities within a constitutionally structured free society. The main objectives are to promote more effective cooperation among law enforcement agencies in the United States and the rest of the free world, to provide an institutional environment that fosters continual personal and professional growth and development, to prepare future law enforcement personnel for leadership roles, and to develop further advanced technical and operational skills.

The philosophy of the national academy is holistic in that a healthy lifestyle and physical fitness are stressed along with academic achievement. A typical session lasts for about 11 weeks. To qualify for the training, one must apply to the FBI and be sponsored by the employing agency. There are no charges for tuition, books, or equipment used. Meals, lodging, and dry cleaning and laundry services are also provided without cost. The individual officers or their agency must assume the travel cost and the assessment fee ($150).

Officers who complete the National Academy Program successfully earn college credit through the University of Virginia. Credits earned are identical to those earned on the main campus at the university. Graduate credits may be earned by those accepted in the graduate courses. All students who complete 13 semester hours of course work with an average grade of C or better and do not fail a course are awarded a certificate by the Division of Continuing Education from the University of Virginia.

The faculty at the academy consists of highly qualified people, with both academic and practical experience. For the most part they are specially trained and educated practitioners who are special agents of the Federal Bureau of Investigation. In addition to the regular staff, some visiting faculty are utilized, including academicians, judges, members of the medical profession, and knowledgeable law enforcement executives.

All students are required to take a minimum of 13 and not more than 19 semester hours of academic work. Usually, a student is required to take one course from each of the following disciplines:

- Forensic science
- Law enforcement communication
- Leadership and management science
- Behavioral science
- Legal issues
- Physical training

Students may petition to drop the required courses in law enforcement communication, forensic sciences, or behavioral science, and upon getting permission, take two courses from one of the remaining disciplines. All students are required to take a 44-hour noncredit fitness in law enforcement course and 30 hours of instruction in the specialized instruction program. Additionally, an enrichment program is available for academy students. This voluntary program is conducted during evening hours and offers students additional, motivational presentations centering on leadership and management issues.

The national academy also offers an advanced leadership course. The leadership challenge is a leadership development program designed to provide students who enroll in the program with practical experience in leadership attitudes and behaviors. The experiences are based on competencies identified in the leadership research. The challenge is designed to provide students with situations that require the student to adopt leadership attitudes and exhibit leadership behaviors. For example, a student may be required to perform volunteer work; to mentor a young law enforcement officer; to tutor a wayward juvenile; to submit a 360-degree evaluation regarding the student's own leadership effectiveness; to read two books on leadership from a selected bibliography; to create a vision for their department; to learn a new computer skill; or to become involved in other practical leadership development exercises. A sample of one course from each training unit is provided below for further discussion (U.S. Department of Justice, 1999).

Forensic Science Unit

- Overview of Forensic Science for Police Administrators and Managers (3 semester hours, undergraduate)

This course provides a basic overview of forensic science and forensic science issues. Instruction includes class lectures, class discussion, occasional demonstrations, actual case reviews, and small-group problems. The course addresses numerous issues, such as managing the crime scene, the role and value of different types of physical evidence, and current trends and issues.

Law Enforcement Communication Unit

- Seminar in Media Relations for the Law Enforcement Executive (3 semester hours, graduate)

This course focuses on contemporary relations between law enforcement and the news media. Particular emphasis is placed on the development of proactive rather than reactive departmental media exposure. More specifically, the seminar challenges the law enforcement executive to learn effective interview techniques, develop organizational messages, either establish or refine departmental press policy, explore fully the role of the departmental public information officer in terms of both internal and external media relations, and examine special issues such as crisis situations, technological changes, and innovative techniques for the development and management of a departmental media training program.

Leadership and Management Science Unit

- Managing Organizational Change and Development (3 semester hours, graduate)

This management course in seminar format is an introduction to organizational development and planned change and focuses on the effect of changes in employees' behavior. Through the use of case studies, the seminar studies the nature and methods of managing, diagnosing, implementing, and evaluating change in law enforcement. The course offers practical suggestions for improving fundamental and large-scale change.

Behavioral Science Unit

- Stress Management in Law Enforcement (3 semester hours, under-graduate)

The course examines the nature of stress in law enforcement and identifies the stressors unique to the profession. A holistic approach to stress management and wellness is introduced to include a wellness profile, the psychology and physiology of stress, critical incident stress, nutrition with regard to cholesterol and stress-related maladies, attitudes, fitness, spirituality, and motivation. The responsibilities of the officer, as well as those of the organization, regarding stress management are offered. Individual stress management/wellness techniques are introduced for officers to maintain or regain control of their lives.

Legal Issues Unit

- Legal Issues for the Police Administrator (3 semester hours, undergraduate)

The course includes a comprehensive examination of the legal problems confronting law enforcement executives. It offers an opportunity to consider legal as well as managerial solutions to problems associated with:

- Hiring, training, promotion, and discipline
- Development of policies regulating employee conduct
- Internal investigations of alleged misconduct
- Development of policies and procedures to reduce potential liability
- Affirmative action
- Recent developments in constitutional criminal procedure

Physical Training Unit

- Fitness in Law Enforcement (44 hours, noncredit)

The course is structured to address fitness as a philosophical base on which fitness and health are enhanced to better serve career goals and objectives. Approximately 20 hours of classroom lecture is combined with 24 hours of personalized physical activity. The course objective is to combine knowledge, discipline, and challenge to produce a positive, proactive lifetime fitness program for each officer and ultimately to have the same carryover to his or her department.

Specialized Instruction Program

- Contemporary Issues in White-Collar Crime (30 hours, noncredit)

This course is designed to provide an officer with an overview of selected contemporary white-collar-crime-related investigative issues. This awareness seminar includes such topics as:

- Money laundering and forfeiture
- Various fraud schemes and activities
- Computer-related investigative issues
- A variety of selected investigative techniques

Northwestern University Traffic Institute

The Traffic Institute was established at Northwestern University in 1936. The institute is a national nonprofit organization that serves public agencies responsible for law enforcement, criminal justice, public safety, traffic management, and highway transportation systems. Local, county, state, and federal government agencies, as well as agencies from foreign countries, are served through programs of specialized training, continuing education, research and development, publications, and direct assistance.

The staff can develop programs designed for instructional requirements of a given agency. The courses are based on an agreement between the agency and the director of the Police Operations and Administration Training Division. If the agency is large enough, with a group of managers/executives who need to be trained on site, the institute can present a custom-tailored seminar at the agency's site. The courses are directed and taught by Traffic Institute professional staff members. The instructors are all former law enforcement officers with an aggregate of over 200 years in enforcement experience. Guest lecturers, including several graduates from the Traffic Institute who supplement the faculty, come from other law enforcement agencies, educational institutions, and national institutions.

A certificate of successful course completion is awarded to each student who fulfills the following requirements:

- At least 90 percent attendance
- Passing grade on a written examination
- Satisfactory class participation

In addition, graduates of the accredited courses may receive two or three semester hours of undergraduate credit from Northwestern University's Division of Continuing Education. In the area of management, the institute offers three main courses (*http://www.northwestern.edu/nucps*).

- *Executive Management Program (EMP).* This program addresses the need for intensive management training that is affordable and takes less time away from the job. The EMP focuses on improving conceptual skills of people in such positions as chief, deputy chief, and similar executive positions.
  - *Duration:* 15 days
  - *Prerequisites:* It is recommended that students be graduates from more extensive programs in police command, such as programs at the FBI National Academy or the Southern Police Institute (SPI), or that he or she has a college degree.
  - *Topics:* community policing, avoiding media brutality, futures, recruitment and selection, and the legal aspects of discipline. Skills taught at the conceptual level include conflict negotiations, leadership for the future, and assessing implications of change.

  Faculty include instructors with an extensive background in teaching management training and conflict resolution; a former chief of police; an award-winning veteran broadcaster with 30 years of experience as a radio and TV news anchorman; an internationally recognized expert in police disciplinary matters; and a noted lecturer on leadership and the future of law enforcement.

- *School of Police Staff and Command.* The School of Police Staff and Command is designed for police managers and their agencies. The course focuses on increasing the knowledge and skills necessary for assuming increased responsibilities in administrative staff or line command positions.

  - *Duration:* Ten weeks

  - *Requirements:* The course is intended for mid- and upper-level supervisory personnel. The students should have at least two years of supervisory experience and be prepared to complete upper-division (i.e., junior and senior level) university coursework.

  - *Topics:* management principles, interpersonal and organizational communication, transformational leadership, organizational behavior, police ethics, managing criminal investigations, current topics in law enforcement management, selection and promotion, labor–management relations, managing internal investigations.

- *Supervision of Police Personnel.* This course is a comprehensive study in first-line supervision and is therefore suited specifically for new managers.

  - *Duration:* Ten days

  - *Topics:* the supervisory challenge, communication, motivational principles, leadership, planning and decision making, responsibilities of a staff officer, cultural diversity, and case studies.

Senior Management Institute for Police

Each year, the Senior Management Institute for Police (SMIP) provides police executives intensive training in the latest management concepts and practices used in business and government. A demanding three-week course, SMIP is taught by faculty from some of the nation's top universities. It is designed for senior police executives who will ultimately lead police agencies throughout the United States and other participating countries. Some exciting changes have taken place at SMIP. In addition to format and materials updating, SMIP's curriculum and faculty have been expanded to include more discussion of the issues that demand the attention of today's forward-thinking law enforcement leaders. SMIP is now held on the campus of the Suffolk University Law School in downtown Boston (*http://www.policeforum.org*).

Purpose. SMIP brings together leading thinkers in corporate and public management to provide intensive training in the best available management theory and practice, innovative solutions to organizational problems, and discussion of important issues in managing public service organizations effectively. The program's goal is to give police managers the same quality of management education available to leaders in other public- and private-sector endeavors.

As a developmental program for the profession's current and future leaders, SMIP focuses on leadership and executive development. The curriculum is much more conceptual than technical, and it requires participants to think in broad terms about their agencies' environments. Cases and class discussions stimulate critical thinking and problem solving. Participants emerge with an understanding of advanced management practices and effective methods of organization and enhanced awareness of the management methods and resources necessary for performing current or future responsibilities. By sharing individual management experiences and exchanging ideas during group discussions, participants gain confidence in their managerial abilities and develop sources of consultation, advice, and support that will endure well beyond the course. The extensive resources of the Police Executive Research Forum and the Senior Management Institute for Police remain available to participants after the course. This commitment has helped make SMIP a national center for the education and training of the future leaders of policing in the United States.

The Course.   SMIP gives participants a clear understanding of general management theory, policy development, planning processes, and organizational structure and behavior. Some of the topics covered include diversity, political management, organizational strategy, performance management, organizational change, leadership, managerial problem solving, labor relations, problem-oriented policing and implementation strategies, process analysis, budgeting, media relations, and new policing strategies and innovations. The program uses the case study method of instruction. Popularized by use in the nation's top business schools, this method combines careful and extensive reading of case materials, including problem analysis and managerial decisions, with classroom discussion of the issues presented in each case. SMIP uses corporate, public, and police agency cases and encourages participants to apply each case's concepts and issues to their organizations. *This is a very demanding, fast-paced, reading-intensive program that requires considerable commitment and hard work in class and after class through independent and group study assignments.* Because of the program's intensity and daily group study, participants are required to stay on site for the program's duration (weekends excepted).

Qualifications.   Enrollment is limited; to qualify for selection one should:

- Be the chief executive or a senior manager in your agency (sworn or civilian) with significant responsibility for major agency activities and have the potential for promotion to chief executive in your current agency or another
- Possess a baccalaureate degree, preferably from an accredited college or university
- Be willing and able to *participate actively* in this valuable management education experience

Nomination Process.   To apply, the candidate's chief must submit a nomination package consisting of a completed application and a letter of nomination from the agency head, specifically outlining:

- The candidate's qualifications and contributions to the department and/or the profession that set the nominee apart from others and demonstrate his or her qualifications for selection to attend SMIP
- The candidate's daily awareness of the demands of this program, including significant reading assignments and required attendance at group study sessions

Federal Law Enforcement Training Center, Management Institute

The Federal Law Enforcement Training Center is located in Glynco, Georgia. With a clearly defined mission to be a provider of world-class law enforcement training in a cost-effective manner, it provides training to over 25,000 students annually. The fee for attending the program is paid by the individual agency and varies every year. Learning methodologies include lectures, small-group discussions, case studies, problem-solving exercises, simulations, role-plays, individual and group presentations, and computer-generated management-style inventories. The FLETC Management Institute designs, develops, coordinates, and administers training programs for federal law enforcement supervisors and managers. The following programs are offered (*http://www.fletc.gov/*).

- *Basic Law Enforcement Supervisor Training Program.* Participants in this program focus on how to apply supervisory knowledge, skills and abilities, such as leadership skills, communications skills, legal responsibilities, stress management, workplace diversity, performance management, briefing skills, and decision-making skills. The staff are all former or current supervisors from various law enforcement agencies. The duration of training is eight class days.
- *Federal Law Enforcement Manager Training Program.* Specifically designed for federal law enforcement managers, this course gives the participants an opportunity to explore alternative management problems and to expand their management tools. Learning methods include participating in problem-solving exercises, small-group discussion, individual and group presentation, role-play, simulations, case studies, short research projects, and computer-generated management style inventories. The duration of training is 10 class days.
- *Situational Leadership Training Program.* This program is a practical approach to management training developed by Ken Blanchard, Patricia Zigormi, and Drea Zigormi. The program teaches that the most effective leaders use a combination of four styles: directing, coaching, supporting, and delegating. Length of training is three class days.

Additional courses offered include:

- Leadership through People Skills Training Program
- Effective Leadership through Understanding Human Behavior Training Program
- Seven Habits of Highly Effective People Training Program
- Customized Programs

Institute of Police Technology and Management

The Institute of Police Technology and Management is located at the University of North Florida in Jacksonville. The institute was established in 1980 and provides management, traffic, and specialty training to municipal, county, state, and federal law enforcement officers. The IPTM trains more than 12,000 students annually. The faculty is composed of professionals who combine law enforcement, technical, and management skills with years of experience and an academic background. It is a full-time staff that comes from government, private industry, and academic environments. The following management courses are offered by the IPTM (*http://www.iptm.org*).

- *Administration, Management, and Supervision of the Field Training Officer Program.* This seminar focuses on the manager and supervisor responsibility for the administration and supervision of the Field Training and Evaluation Program. The program is designed for veteran and newly assigned FTO supervisors. Topics include the supervisor/manager's FTO program responsibilities, FTO selection, retention, and evaluation, burnout, incentives, and liability issues. Length: three days.
- *Comprehensive Police Fleet Management.* This course focuses on vehicle testing, records management, and other crucial aspects of police fleet management. A workshop on microcomputers is built into the program. Topics include management issues and skills and risk management. Length: five days.
- *Developing Law Enforcement Managers.* This course focuses on problems that police executives encounter, such as project management, policy development, leadership and motivation, police budgeting and cost control, communication, time management, and decision making. The course has been designed specially for top administrators from agencies of any size and for those middle managers who are on the way up the executive ladder. Length: five days.
- *Field Training Program for Communication Officers.* This seminar is designed for the communications dispatcher training officer and those who immediately supervise him or her. The main focus is on the training officer's role in evaluation, teaching, and remedial training, and the

training case study review, trainee termination procedures, teaching principles and methodology, remedial training and narrative reports, and legally defensible performance evaluation. Length: three days.

- *First-Line Supervision Course*. This course is created for the new supervisor and assists with understanding the transition from operations to management responsibilities. The participant will learn communications and semantics, management theory, organization theory, human relations, and other important topics that are important for the new supervisor. Length: two weeks.

- *Implementing and Managing Community-Oriented Policing*. This course is designed for the person responsible for implementation of a department's community policing effort. The program focuses on topics such as program management, dealing with community groups, and development of survey instruments. Length: five days.

- *Leading Law Enforcement into the Twenty-First Century*. This course focuses on knowledge and skills that will be required to lead law enforcement agencies in the near future. Topics include: management—where do we go from here?, flexibility in management, leadership versus management, and developing and facilitating effective teams. Length: five days.

- *Management of the K-9 Unit*. This course is designed for supervisors in the police service dog unit. The training focuses on organization and management and dog knowledge. Length: five days.

- *Managing Criminal Investigators and Investigations*. This is a seminar for new and experienced investigative supervisors. The participants review and discuss practical managerial and leadership principles which are intended to resolve the typical time management difficulties encountered by investigative supervisors. The classes include skills and attributes of investigators, selecting officers for investigative positions, supervising investigators, techniques for case screening, effective case management, and case review techniques. Length: five days.

- *Police Command and Staff*. This is a seminar for the middle manager in the agency that wants to receive high-quality, state-of-the-art training but lacks the time to put into an extensive course. The course consists of a review of management theory and principles. In addition, there is a series of traditional management topics, including decision making, delegation, leadership, and motivation. Length: four weeks.

- *Police Discipline Seminar*. This seminar is for the police executive, administrator or supervisor who has responsibility for overall discipline in an agency. Some of the topics are a definition of police discipline, the law and discipline, and records training and discipline. Length: three days.

- *Police Traffic Management*. This is a course that provides supervising and midlevel management law enforcement personnel with the latest traffic technology. Two weeks are dedicated to management subjects. Length: four weeks.

- *Seminar for the Field Training Officer*. This is a program for the new field training officer (FTO) and those who supervise him or her. The main focus of the program is the FTO's role in evaluation, teaching, remedial training, and the effects of his or her style and beliefs on recruit growth and achievement. Length: five days.
- *Workshop for Police Traffic Commander*. This course focuses on exchanging information between people who command traffic divisions or units. Length: five days.

Operation Bootstrap

Founded in 1985 by Chief Michael Shanahan of the University of Washington Police Department in Seattle, the program focuses on management training for law enforcement professionals. The name of the program comes from "pulling police managers up by their bootstraps." Now termed the Corporate Training Alliance Program, 1300 law enforcement agencies are registered as members. Cost of the program is $175 to $350 a day per course, but the program is tuition-free. There is also an additional $100 administrative fee. The following courses are offered.

- *Maximizing Individual Effectiveness*. This seminar focuses on making the individual assess his or her influence on groups. Techniques such as problem-solving classes, feedback on your conflict management, risk taking, stress management and interpersonal communication behaviors, small-group simulations, and role-plays are used.
- *Facilitation Workshop*. This workshop focuses on improving facilitation skills, managing groups, and effective questioning techniques.
- *Managing Conflict*. This course focuses on the investigation of factors that create conflicts, how to confront the problems, how to resolve the problem, developing the coping skills, how to resolve business problems, and how to resolve a current conflict situation.
- *Presentation Skills Workshop*. The student/participant has to present at least five speeches and see oneself on videotape. Feedback focuses on performance, stage fright, and effective presentation techniques.
- *Communication Workshop*. Participants in this seminar focus on the practice of multiple listening and observation skills and how to use a standard model for achieving the skills.
- *Managing People and Performance*. The course focuses on how to identify and emphasize key managerial responsibilities. Feedback is given on each person's management practices and communication style. In addition, the course looks at motivational needs, leadership behavior, case discussions, and how to develop effective plans for successful personal and organizational effectiveness.

- *Product Management Concepts.* The course offers training to product managers in regard to framework, models, and tactics for dealing with the functional areas in law enforcement agencies.

Bill Blackwood Law Enforcement Management Institute of Texas

The Law Enforcement Management Institute of Texas (LEMIT) is the largest statewide preparation program for police management in the United States. The program was introduced in 1989 and in 1993 became a part of Sam Houston State University in Huntsville, Texas. The mission of LEMIT is to provide exceptional education, training, research, and service to customers in the law enforcement community, in order to inspire excellence in leadership and to enhance the effective delivery of law enforcement services.

LEMIT offers several programs in management training and research programs. The programs are categorized into eight areas (*http://www.shsu.edu/lemit*):

- Command Staff Leadership Services
- Executive Issues Seminar Series
- Graduate Management Institute
- New Chief Development Program
- Professional Conference Support
- Special Programs
- TELEMASP
- Texas Police Chief Leadership Series

The Graduate Management Institute (GMI) is the most comprehensive program offered by LEMIT. The main purpose of this course is to give police executives the knowledge and skills necessary for successful leadership in today's law enforcement agency. The length of the program is 18 days of instructional modules, and it includes selected reading assignments and a comprehensive policy research project. The curriculum consists of three main areas:

- *Module I:* General Management Principles
- *Module II:* Political, Legal, and Social Environment of Policing
- *Module III:* Law Enforcement Administration and Leadership

Texas Regional Community Policing Institute

The Texas Regional Community Policing Institute (TRCPI) is located at Sam Houston State University and provides training for criminal justice agencies and community groups throughout Texas. The program was established to serve as a resource for Texas law enforcement agencies by providing information, training materials, and other support for law enforcement professionals who seek to adopt more of a community orientation in their policing activities. More than 5000 students have received training at TRCPI. The following advanced training programs are offered (*http://www.shsu.edu/cjcenter/trcpi/trcpi.htm*):

- *Executive Leadership Series.* This course is designed for top management and focuses on community policing philosophy and how to get the community involved in community policing. This seminar is limited to 25 participants.
- *Community-Oriented Policing and Problem Solving (COPPS).* This program is designed for line officers and middle managers and is basically an introduction to community- and problem-oriented policing. The course is limited to 30 participants.
- *Executive Issues Seminar Series.* This program is limited to agency top command staff and addresses emerging issues in policing. The participants are directly involved in the sessions, which makes the course open to only 20 participants.

Police Training Institute

The Police Training Institute at the University of Illinois has trained over 57,000 law enforcement officers since its founding in 1955. The mission of the institute is to provide needed training to criminal justice personnel, both preservice and in-service, at basic and advanced levels. The training should be of highest quality and responsive to changes in criminal justice.

The advanced training program offered by the institute is composed of two training modules, titled the Integrated Training Management Program. This program includes the Field Training Officer Course and the Leadership Development for Supervisors Course. The main purpose of both of these programs is to provide students with improved knowledge in tactical patrol operations (*http://www.pti.uiuc.edu*).

- *Field Training Officer Course:* designed to improve knowledge in team building, leadership, and communication. The length of the program is 12 weeks, in which the participants will have the chance to observe, evaluate and recruit officers, and thereafter discuss their observations and evaluations.
- *Leadership Development for Supervisors Course:* lasts only 32 hours and is a hands-on course. In this program, supervisors experience problems in communication, leadership, problem solving, and general human resource development, and they learn how to cope with them. Topics include:
  - The Supervisor as Leader
  - Identifying and Applying Leadership Styles
  - Developing Effective Communication Skills

- Active Listening
- Problem Solving
- Supervising Tactical Operations
- Human Resource Development

SERVANTS OF TWO MASTERS: THROUGH A DILETTANTE OR TOTAL QUALITY MANAGEMENT?

Similar to the concepts defined in Chapter 8, the definition of supervisory/executive/management training becomes the key to understanding what, how, and why is offered underneath the megalabel "advanced training." The available literature on management of organizations and organizational behavior offers almost as many definitions of management as there are writers in the field. The review of some of the programs offered to supervisor/managers/executives in the field of law enforcement provides an illustration of a similar situation. Numerous programs offered throughout the country range in their approach from very thorough and intense, like the one offered by the FBI Academy, the Northern Traffic Institute, or SMIP, to a more nonchalant or much more limited approach, such as like the three- or five-days courses offered by the IPTM , PTI, or LEMIT (it must be noted that the three also offer longer training modules).

The main problem, however, is not only the length or duration of a given course, but who offers it, why it is offered, and what is being taught. The "who," "why," and "what" of law enforcement management training can be better understood by first analyzing the concept of management or executive goals and objectives in other, non–law enforcement organizations.

The training of police personnel is an expensive venture to begin with; advanced/management training is an even more expensive proposition, since the department not only has to pay for the training itself but has to provide the salary for the officers who participate in the training programs. The longer the program, the more costly it is for a given department. However, many police departments continue to spend their training dollars without attempting to discover exactly what type of training is desirable. Chruden and Sherman (1984) advocated the use of task analysis to ensure the relevance of training, analysis that would provide a detailed examination of a job and the role played by the employee in that job. According to this approach, once the role, tasks, and job have been clearly defined, training needs should be fairly easy to assess. Utilizing this concept, one should easily be able to answer the questions of "what" should be offered and "why." However, the remaining question of "who" should offer the training remains unanswered, despite utilization of the task analysis concept.

Prior to discussing possible answers to the "who" problem, the author would like to challenge the concept of task analysis as a feasible tool for assessing the management training needs in law enforcement. Going back to our earlier concept of police organization being the servant of, at least, two masters, the utilization of task analysis becomes slightly problematic. A task analysis is a detailed examination of a job and the role played by the employee in that job. If one agrees with the idea of the servant of two masters, the management position in any given police organizations becomes ambivalent in nature, since police executives will have to answer not only to the public but also to the politicians. Therefore, their task is a complex one, not easily defined, and the role they play is virtually elusive, since in essence they do have to serve both masters.

A common thread that appears in the various definitions of the concept of management behavior is the manager's requirement to accomplish organizational goals or objectives (Hersey and Blanchard, 1993). This definition of managerial role in any for-profit organization, and even in some nonprofit organizations, is a clearly understandable one. Any organization, except for a police organization, will strive in a legitimate and sometimes illegitimate manner, to accomplish the organizational goals and objectives. All organizations operate based on two sets of goals: formal and informal (Green, 1999). The formal goal of any organization is the one that the organization "sells" to the public; the informal goal is the goal that only the employees within the organization (and not all the employees) are aware of. Middle and upper management are clearly aware of that informal goal. Police organizations profess their commitment to community policing and serving and protecting the public. Their formal goal is clearly defined in the media through short indoctrinations of the officers, during community-oriented training, and so on. However, the informal goal, the bottom line, of satisfying the needs of the political entities involved in their operational decisions is rarely, if ever, acknowledged. Therefore, the problem of "what" and "how" remains unresolved. The more intensive and in-depth courses for police supervisors and managers, courses that emphasize conceptual skills, are not necessarily "desirable" for police executive personnel. The shorter and shallow approach of four, five, and eight days of training appears to be much more "valuable" to the organizational goal of satisfying two masters.

In their seminal work *Management of Organizational Behavior*, Hersey and Blanchard (1993) define management as "the process of working with and through individuals and groups and other resources to accomplish organizational goals" (p. 5). They further emphasize the distinction between management and leadership, pointing to the fact that management is a special kind of leadership in which the achievement of organizational goal overrides everything else.

In leading and influencing, Hersey and Blanchard (1993) identified three general skills or competencies, cognitive, behavioral, and process:

1. *Diagnosing:* understanding the situation and knowing what can reasonably be expected in the future
2. *Adapting:* adapting behaviors and other resources to close the gap between current achievements and what can be achieved
3. *Communicating:* communicating effectively in order to achieve the desired result

In further expanding on the issue of "what" and "how," the leading and influencing aspects of managerial skills seem not only to be neglected in most advanced training modules, but in some instances are encouraged to be ignored. The prime example is training at the FBI's National Academy which allows students to petition to drop the law enforcement communication and behavioral science modules and substitute modules with courses from other areas. On the other hand, the academy requires 44 hours of fitness training. It has to be noted that the FBI academy has put together a very intense and up-to-date training module, which should be replicated by some larger in-house academies. But missing from the program is a mandated emphasis on the three principal competencies of managers: cognitive, behavioral, and process. All three could be be enhanced by mandating courses in the law enforcement communication and behavioral sciences modules instead of allowing the students to substitute for them with other courses.

A different approach to advanced police training was offered by Feltes (1999), who emphasized that the main objective of police management training is not to train police officers as management experts but to give them:

- An initiation in the practice of modern management
- Insight and practical understanding of the techniques and tools available
- The ability to identify possible benefits and opportunities and to apply the tools for the accomplishment of their daily tasks
- The possibility to manage decisions in a structured way and to run an operation effectively and efficiently

Feltes's approach appears to be a combination of Hershey and Blanchard's recommendations and the author's belief that management training is being handled dilettantely rather than from a total quality management approach. Feltes identified three main topics to be addressed in management training:

- The public as client; the fundamentals of public service, with emphasis on the notion of service
- Attitude and communication training, conversation, body talk, effect of uniform
- Behavior in various practical situations

Furthermore, he recognizes the absolute need for officers to acquire managerial capabilities and advocates for the inclusion of management training as an integral part of basic training, rather than being instructed by means of refresher courses or seminars, which should be used only as part of a continuous updating program. Feltes's view of management training is refreshing and provides a partial answer to the "what?" and "how?" dilemma. However, "who?" remains wide open. The review of training modules offered in this chapter emphasized the importance of the professional qualifications of the experts who deliver the training as well as the credentials of the institutions and institutes that provide the training.

Total Quality Management

The idea of total quality management was adapted to law enforcement environments from the private for-profit sector. The customization of the concept for the realities of life in police organizations was never factored in. If it was, it is certainly very well hidden. To understand more comprehensively the problems related to who should deliver supervisory/management training to police executives, it will be beneficial to review the basic concepts of the total quality management. Following are the 10 basic values of TQM (Hodgetts, 1996, pp. 8–10):

1. *Customer-driven quality*. Quality is judged by the customer, so all product and service efforts are designed to contribute value to the customer and lead to buyer satisfaction.
2. *Leadership*. Clear and visible quality values must be set.
3. *Continuous improvement*.
4. *Association between participation and development*. Allocate funds for associate education and training opportunities.
5. *Fast response* (to customer needs). Maintain time and quality standards, but simplify work procedures and processes.
6. *Design quality and problem prevention*. Prevent problems in products and services from the outset so that they don't have to be corrected later.
7. *Long-range outlook*.
8. *Management by fact*. Use objective data (e.g., customer surveys), not stories or comments from customers.
9. *Partnership development*. Develop a spirit of partnership both internally among employees and externally with customers and other organizations
10. *Corporate responsibility* (e.g., business ethics, protection of public health) and citizenship (leadership, support of publicly important purposes, partnerships with outside agencies to partner for these purposes).

The areas of main importance in TQM are (p. 11):

1. Customer and employee satisfaction
2. Data collection
3. Employee training
4. Design of the structure
5. Provision of recognition and rewards
6. Development of a system of continuous improvement

The TQM approach was developed further by introducing the concept of reengineering, which was also emphasized in various advanced training modules as another practical approach to the management of police organizations. Hammer and Stanton (1995) provided a thorough definition of the reengineering concept: "Reengineering means, fundamentally, to rebuild the company on the customer's behalf. This requires a deep understanding of the customer needs. . . .The reengineering leader must orient the customer through explicit communications, targeted reward systems, and by getting everyone to talk to customers" (p. 97). Following are the characteristics required for successful reengineering leadership, organizational readiness, and style of implementation identified by Hammer and Stanton (1995, pp. 92–99).

1. There must be a senior executive who has a strong commitment and a true understanding to reengineering his or her organization and the authority to make the organizational changes needed for successful reengineering to actually occur.

2. The leader must communicate a clear and simple but persuasive vision of the reengineering effort that employees can rally behind.

3. The leader must contribute to his or her own energies and hard work to the effort. He or she cannot delegate the work involved, although employing other senior-level people is necessary to utilize their talent and also to show the seriousness of the reengineering effort. Furthermore, all senior-level managers must agree with the values of the effort or it will not succeed.

4. Explain to employees why the current organizational atmosphere must change. Discourage any complacency that nothing is wrong with the organization (so why change it?). Point out that one never knows what will happen in the future.

5. Employees must understand reengineering. The more they don't understand it, the more they will be suspicious of it as the latest management fad or euphemism for downsizing.

6. If painful layoffs or similar negative experiences have occurred within the organization, emphasize to employees that management made mistakes in the past. Make sure that you bring quick positive results for employees to show you are for real.

7. Make sure that you use your top-rated employees in the reengineering effort. You need people who aren't afraid to break old rules. The suggestions of bold rule breakers should be applauded, while timid suggestions should be rejected summarily. Outsource if people are unwilling to break from the past.

8. Do not confuse TQM with reengineering.

9. TQM and reengineering share many characteristics: focus on customers, orientation toward processes, and a commitment to improved performance.

10. There are important differences between TQM and reengineering:

TQM	**Reengineering**
Incremental improvement through structured problem-solving	Radical improvement through total process redesign
Assumes that the underlying process is sound and seeks to improve it	Assumes that the underlying process is not sound and seeks to replace it

11. Reengineering can be seen as the next step after TQM.

Milakovich (1995) provided some additional insights into implementation of the TQM process in organizations by improving service quality:

1. Make customer satisfaction the primary goal and ultimate measure of service quality.
2. Broaden the definition of "customer" to include both those internal to the organization (e.g., employees in other departments) and those external to it (e.g., vendors, taxpayers, contractors, regulators, suppliers).
3. Develop a common vision of the mission of the organization based on extended customer requirements.
4. Communicate a long-term commitment to all customers, reward teamwork, and encourage process improvement and innovation efforts at all levels.
5. Provide expanded education, training, and self-improvement opportunities in supervisory and leadership skills in order to exceed valid customer requirements.
6. Ensure individual involvement by establishing and supporting organization-wide process improvement teams.
7. Recognize, support, and acknowledge employee loyalty, trust, and team participation.
8. Eliminate fear in work and remove barriers to developing pride in service (i.e., empowerment).
9. Provide the proper tools and training for everyone to respond to extended customer requirements.
10. Make the necessary changes in public organizations to implement the preceding goals successfully.

The rather extensive review of the TQM, TQS, and reengineering concepts becomes necessary in light of the fact that these concepts became not only the hottest terms in public management in general but also have been heavily endorsed for adoption by police agencies during the 1990s and into the future. Goldstein (1993) found it extremely troubling that most police agencies do not invest enough in explaining to their senior officers why the TQM change is necessary, and why the agencies do not provide their supervisors and managers with the freedom required for them to act in their new role.

The answer to Goldstein's plea and to the "who?" question (who is supposed to deliver the advanced/supervisory training?) can be found in a brief analysis of the concepts of total quality management. Police agencies do not invest enough in explaining to their senior officers why the TQM change is necessary because the way the TQM change is presented cannot be implemented in police organizations. The theory behind the concept is enlightening and certainly workable for implementation in for-profit organizations that cater to one master—the recipient of their product. Police organizations cater to two masters, the politician and the public, and furthermore, the public is composed of many masters—each interest group that exercises pressure on a given police department. It is virtually impossible to create a common vision of a department based on extended customer requirements since the customers are too demanding and too numerous.

Almost every point of the TQM philosophy can be addressed by describing the obstacles to its implementation in a law enforcement environment. We will address the 10 principles of the TQM philosophy by pointing to some of the obstacles.

1. One customer's satisfaction is another customer's dissatisfaction (what is loud music for one is barely a minor noise for another).
2. There are not yet enough leaders in law enforcement to set clear and visible quality values.
3. Improvement is conditioned upon resources that are for the most part controlled by external (to police organization) environments.
4. Resources for training and education are scarce at best, if not nonexistent, especially in smaller police departments.
5. In law enforcement environments, fast response to customer need does not guarantee satisfaction.
6. Prevention of problems such as drugs, gambling, and prostitution is, at best, limited in nature for police organizations. Other quality-of-life problems are either problematic in nature (bothering some but not others) or can be eradicated by only aggressive enforcement.
7. Police work is still reactive in nature, despite the drive toward a proactive/preventive approach; therefore, the long-range outlook is limited.
8. Citizen satisfaction surveys are of limited applicability, especially in larger cities and high population mobility, but also depend on a host of factors unrelated to overall police performance (e.g., the most recent experience/interaction with law enforcement can skew the answer considerably).
9. Community-oriented and problem-solving policing is based essentially on the concept of partnership with organizations, but organizations represent various interest groups, and the partnership is really limited and frequently superficial.
10. Corporate responsibility for police organizations is really an issue of how well a given organization can balance responsibility among its many masters. The answer to how well this responsibility is balanced today is still problematic.

The 10 points in this response to the feasibility of implementation of TQM principles in law enforcement environments provide the answer to the question of who should teach supervisory/management courses designed for police practitioners. It is imperative to utilize instructors who have practical experience in the field of law enforcement and understand the necessity to set up a decision-making process that is not based on the conceptual tenets of the TQM philosophy. The brightest stars in the field of management cannot convey this message to law enforcement practitioners; they can introduce the concepts but cannot provide guidelines for practical implementation. Implementation technique is what a police supervisor/manager is looking for in advanced training. The old adage of "take the politics out of policing and the police out of politics" is still in the forefront of management of organizational behavior in any law enforcement agency. The sooner we realize and acknowledge this fact, the better our understanding of the structure and volume (or lack of volume) of police supervisory and management training will become.

SUMMARY

This chapter follows the structure of Chapter 11, which depicted specialized and specialist training rather grimly. Our detailed description of management training in this chapter has been similar. Although the importance of effective management and supervision has always been emphasized by modern police organizations, at least rhetorically, the practical side leaves much to be desired.

The instructional modules reviewed in the chapter vary from extremely effective programs of high quality such as those offered by the FBI's National Academy or the SMIP Institute (which despite their overall quality are somewhat problematic with regard to relevancy) to the modules offered over the Internet, some of which are of rather dubious quality.

Additional problems appear to be the scarcity of advanced management programs and relative lack of consistency on the part of many agencies in exposing their supervisory staff to any type of advanced training. The reluctance to invest time, money, and other resources in training the supervisory cadre is inconsistent with the move toward implementation of some of the principles of the community-oriented policing. The recommendations included at the end point to the need for high-quality programs offered by quality instructors in high-quality institutions to a broad range of police supervisors.

REFERENCES

Bill Blackwood Law Enforcement Management Institute of Texas. Retrieved September 5, 2000 from the World Wide Web: *http://www.shsu.edu/lemit*.

Chruden, H. J., and Sherman, A. W. (1984). *Managing Human Resources*, 7th ed. Cincinnati, OH: South-Western.

Federal Law Enforcement Training Center. Retrieved September 5, 2000 from the World Wide Web: *http://www.fletc.gov*.

Feltes, T. (1999). Improving the training system of police officials: problems of creating an international standard for police officers in a democratic society. Unpublished paper. University of Applied Police Sciences, Villingen-Schwenningen, Germany.

Goldstein, H. (1993). The new policing: confronting complexity. Washington, DC: U.S. Department of Justice, Office of Justice Programs, National Institute of Justice.

Green, G. S. (1999). *Occupational Crime.* Chicago, IL: Nelson-Hall.

Hammer, M., and Stanton, S. (1995). *The Reengineering Revolution: A Handbook.* New York: Harper Business.

Hersey, P., and Blanchard, K. H. (1993). *Management of Organizational Behavior: Utilizing Human Resources,* 6th ed. Upper Saddle River, NJ: Prentice Hall.

Hodgetts, R. M. (1996). *Implementing TQM in Small and Medium-Sized Organizations: A Step-by-Step Guide.* New York: Amacom.

Institute of Police Technology and Management. Retrieved September 5, 2000 from the World Wide Web: *http://www.iptm.org.*

Milakovich, M. E. (1995). *Improving Service Quality: Achieving High Performance in the Public and Private Sectors.* Delray Beach, FL: St. Lucie Press.

National Advisory Commission on Criminal Justice Standards and Goals (1973). *Police.* Washington, DC: National Institute of Law and Criminal Justice.

Northwestern University Traffic Institute. Retrieved September 5, 2000 from the World Wide Web: *http://www.northwestern.edu/nucps.*

Police Training Institute. Retrieved September 5, 2000 from the World Wide Web: *http://www.pti.uiuc.edu.*

Senior Management Institute for Police. Retrieved September 5, 2000 from the World Wide Web: *http://www.policeforum.org.*

Texas Regional Community Policing Institute. Retrieved September 5, 2000 from the World Wide Web: *http://www.shsu.edu/cjcenter/trcpi/trcpi.htm.*

Thibault, E. A., Lynch, L. M., and McBride, R. B. (1998). *Proactive Police Management,* 2nd ed. Upper Saddle River, NJ: Prentice Hall.

U.S. Department of Justice (1999). *FBI National Academy General Instructions 199th Session,* October 3–December 17. Washington, DC: Federal Bureau of Investigation.

LIABILITY AND TRAINING
Familiarity Breeds Contempt

INTRODUCTION

In the last decade of the twentieth century, much attention was directed toward the wrongdoing of some police officials. The infamous cases of alleged police misconduct in the O. J. Simpson case, the shooting of Amadou Diallo, the sodomizing of Abner Louima, and a number of cases involving what has been termed as *racial profiling* are just a few examples of incidents in which the effectiveness of police training was raised, among other accusations of less than professional police response.

This chapter provides a theoretical framework for understanding the critical issues involved in police officers' exposure to potential liability and the disastrous outcomes this can bring to the individual officer, the organization, and the entire profession of policing. From the theoretical to the practical, an overview of liability-related cases is offered, together with state and federal laws and regulations that serve as a basis for libel suits.

VICTIMS AND VICTIMIZERS

Even though the numbers quoted are more frequently anecdotal than empirically collected, it appears that medical doctors are involved annually in thousands, if not hundred of thousands, of medical errors and mistakes, commonly referred to as *malpractice*. In comparison to the "malpractice" incidents of police officers, the numbers are certainly staggering. However, the public does not seem to engage in outbreaks of rage on the streets of our cities condemning individual doctors, hospitals, or the medical profession of medicine, even though several recent cases have been truly outrageous.

Why is the public in general so much more tolerant of poor behavior by the medical profession than by the police? The answer seems to be related, at least partially, to the topic of this book—training and education. Both professions deal with our most precious possession—our life. The medical profession treats this responsibility with utmost seriousness by demanding that its practitioners undergo an extremely extensive training and education period, before allowing them to practice. At the other end of the spectrum, the police profession allows its practitioners to undergo less than minimal training and sends them to the field to be engaged in activities that also involve daily dealings with human life.

Of course, the comparison might seem simplistic on the surface; critics of the comparison might denounce it as inappropriate and even insulting. But the day a young doctor begins to practice medicine does not differ very much from the first day on duty of a police officer. They both may face issues of life and death, but the doctor has long and extensive years of education and training at his or her disposal, whereas the officer's educational and training record are no match and offer little protection.

For the general public, the difference between the education and training of a medical doctor and a police officer is quite obvious. The "excuse" for medical error is almost built in to the medical profession. If a doctor commits a malicious act, he or she must be insane, crazy or deserves pity, even from the victim, and of course does not represent the medical profession. On the other hand, an officer who brutalizes another human being represents not only his deranged self but first and foremost, the entire police organization.

Physicians usually carry very expensive malpractice insurance. Police officers are frequently covered by their employing agency or by the municipality. However, as mentioned in Chapter 7, they are not covered for the emotional damage one case causes not only to the individual but to the entire agency and its employees, who often have to face crowds accusing them of brutality, racism, sadism, and cruelty.

It is easy for the public to feel anger and resentment toward the entire police profession since the public is not aware that the people they are accusing did not receive adequate education and training that would enable them to carry the label of true professional.

This chapter is dedicated to the skeptics who feel that the need to establish professional, extensive, and far-reaching standards for police training is just a mirage that has no place in American law enforcement in the United States. Those who feel that establishing some standards for police training is against our traditions and that our police officers do not deserve our best efforts to establish a true police profession, are going to watch our society pay the ultimate price—people of lesser and lesser quality are going to be attracted to the profession. Effective policing is based on qualified personnel; without it, professional law enforcement as we would like to envision it will not be possible. Furthermore, it is crucial to recognize that even though Berkley has claimed that "democracy is hard on policing," the more imperative point to be made, following Reith's approach to policing, is that without effective professional policing, there will be no democracy.

To illustrate the gravity of the situation, a brief review of the basis for police liability, followed by a number of number of cases and studies pertaining to police liability and training, are presented and discussed, together with some recommendations for the future. In his extensive study of police liability (which should become mandatory reading for police officers in any basic training academy), Kappeler (1997) summarizes the legal basis for establishing civil liability. The following description of state tort law is adapted from his work. In general, there are two ways to file a suit against municipal police misconduct. It can be done either by filing a lawsuit in a state court claiming negligence or intentional failure to perform a police duty in violation of state law, or through a civil suit brought in a federal court by a plaintiff who claims that the police violated either a constitutional or a federally protected statutory right (Kappeler, 1997, p. 17).

STATE TORT LIABILITY

The first type of lawsuit, filled in state court is brought under state tort law. Kappeler (1997) identifies three commonly used types of tort law under state law: (1) strict liability tort, (2) intentional tort, and (3) negligence tort.

Strict liability torts are usually associated with behaviors that are extremely dangerous, and any person engaging in such behaviors can be substantially certain that his or her conduct will result in injury or damage. These torts do not usually apply to police officers, with some exceptions when certain law enforcement policies allow for use of military or military-style tactics to control drug trafficking, or use of chemical agents to control illegal drug crops.

Intentional torts require the plaintiff to prove that an officer's behavior was intentional and that this intentional conduct led to damage or injury. This requirement does not mean that the officer "intended" to inflict damage or injury but rather that the officer intended to engage in the conduct that led to the injury. The intentional tort assumes elements of knowledge and foreseeability of danger associated with engaging in particular behavior and the extent to which an officer's behavior was intentional. There are a number of forms of intentional torts that can be brought against police officers. Some of them are reviewed briefly below.

- *Wrongful death.* Determination of wrongful death is based on state statues. These cases are often brought by the family of the deceased and involve claims of improper use of force that resulted in the death of the citizen or failure of the officer to prevent death, while the officer can be held liable for wrongful death. Depending on the officer's behavior and state law, punitive damages can be awarded against the officer and the police department.

- *Assault and battery.* Assault requires that an officer act in a manner that causes injury or places a person in fear of harm. To establish liability, it is sufficient to show that the officer caused a person to fear for his or her safety. Battery requires offensive or harmful contact between two persons. This conduct does not require that the officer intended to inflict harm or that the officer realized the conduct to be offensive to a person. It needs to establish that the officer acted without the consent of the party and made some sort of offensive contact.

- *False arrest and imprisonment.* False arrest refers to unlawful seizure and detention of a person's liberty without his or her consent. It is not necessary for the officer to restrain the person physically. It needs to establish that a reasonable person under similar circumstances would conclude that he or she was no longer free to go about his or her activities. However, the plaintiff must show several factors to establish police liability:
  - The police willfully detained them.
  - Their detention was against their will.
  - The detention was made without the authority of law.

  False imprisonment is similar to, but distinguished from, false arrest in that an officer may have had probable cause to arrest but may still be found liable for false imprisonment. The plaintiff must show several factors to establish police liability:
  - An intent to confine.
  - Acts resulting in confinement.
  - Knowledge of the confinement or harm.

  Plaintiffs are usually successful in establishing police liability where police officers:
  - Fail to follow proper booking procedures.
  - Prevent defendants from being properly arraigned.
  - Restrict a defendant's access to court.
  - Improperly file criminal charges against suspects.

Negligence tort refers to inadvertent behavior that results in damage or injury. Negligence requires a lower level of foreseeability of danger than does intentional tort. The standard applied in negligence tort is whether an officer's act or failure to act created an unreasonable risk to another member of society. Four elements are required to establish a case of negligence. Each of these elements must be shown before a police officer can be found liable for a claim of negligence. The elements include:

1. *Legal duty.* Covers behaviors recognized by the courts which require police officers either to take action or to refrain from taking action in particular situations.
2. *Breach of duty.* The plaintiff must establish that an officer breached a duty to a citizen.

3. *Proximate cause.* The plaintiff must prove that an officer's conduct was the proximate cause of injury or damage.

4. *Damage or injury.* The plaintiff must show actual damage or harm and that the damage was such that it substantially interfered with an interest of a person or his or her property.

Following are some of the most frequent forms of negligence brought against the police:

- Negligent operation of emergency vehicles
- Negligent failure to protect
- Negligent failure to arrest
- Negligent failure to render assistance

FEDERAL LIABILITY LAW

Section 1983 of the United States Code is the vehicle used most frequently by citizens to challenge state or local officials who they allege deprived them of their constitutional rights. This federal statutory remedy was derived from the Civil Rights Act of 1871 and was originally passed to provide a mechanism for eliminating Ku Klux Klan activity in the South by providing a neutral forum for newly freed slaves to bring claims against government officials who violated their rights. This legislation has a number of sections that allow citizens to bring civil actions against the police. Despite the fact that criminal charges can be brought against the police for violating a citizen's constitutional rights, litigation under the civil sections of the legislation is far more frequent.

Currently, the legislation allows persons whose civil rights are violated by government officials acting under the color of state law to bring civil suits in federal court to recover from damages or injuries that reach a constitutional level of deprivation. The court's interpretation of the Civil Rights Act of 1871 allows persons who have had their constitutional rights violated by the police to bring civil suits against law enforcement officers, police agencies, and municipal government. This legislation has a number of stipulations:

1. The plaintiff must be a protected person within the meaning of the legislation.
2. The defendant must also be a person within the meaning of the legislation.
3. Citizens seeking redress must show that an officer was acting under the color of state law.
4. Citizens seeking redress must show that the alleged violation was of constitutional or federally protected right.

Unless conditions 3 and 4 are both established, liability is barred under the provisions of the legislation; however, civil suits may still be brought under tort actions (Kappeler et al., 1993).

Acting under the Color of Law

Defendants in a Section 1983 action must be acting under the color of state law to be held liable for constitutional violations. Acts of private individuals, unless they conspire with state actors or the acts of law enforcement officers not related to employment, usually are not actionable under the provisions of this legislation. Off-duty police officers employed on second jobs can also be found liable as acting under the color of law if they perform police functions (Kappeler, 1997).

In their study of cases of police civil liability while acting under the color of law, Vaughn and Coones (1995) conclude that several courts hold that all actions of police officers are conducted under color of law if a departmental policy mandates that officers are always on duty. These courts expressed an opinion that certain departmental policies clearly define all officer actions as official and under color of law. Given these legal realities, if departments have policies that require off-duty officers to be armed, the departments must take appropriate steps to ensure that the officers are properly trained. The more activities a given department defines as nonpolice business and the more training that officers receive in how to act when off duty and when moonlighting, the less the chance of an officer being characterized as under color of law.

Vaughn and Coomes compiled a very helpful list of factors that a police trainer should definitely look at when devising liability-related training. They identified the specific factors that courts consider in determining whether a law enforcement officer acts or does not act under the color of law. The factors are divided into two broad areas: (1) factors that lead courts to determine that the police were acting under color of law and (2) factors that lead courts to determine that the police were not acting under color of law. Following are factors in both categories (Vaughn and Coomes, 1995, p. 409):

- Under color of law
  - If officers identify themselves as law enforcement agents
  - If officers perform duties of a criminal investigation
  - If officers file official police documents
  - If officers attempt or make an arrest
  - If officers invoke their police powers outside their lawful jurisdiction
  - If officers settle a personal vendetta using police power
  - If officers display or use police weapons or equipment
  - If officers act pursuant to a state statue or city ordinance
  - If a departmental policy mandates that officers are always on duty
  - If officers intimidate citizens from exercising their rights
  - If the department supports, facilitates, or encourages off-duty employment of its officers as private security personnel

- Not under color of law:
  - If officer's inaction does not constitute state action
  - If officers commit crimes in a personal dispute without invoking police power
  - If officers act as federal agents
  - If officers report the details of alleged crimes as private citizens
  - If officers work for a private security company and do not identify themselves as law enforcement personnel
  - If officers do not invoke police powers
  - If departments remove the officers' lawful authority

In cases involving federal officials, the actions are tested under the Bill of Rights and are known as *Bivens' actions*, referring to *Bivens v. Six Unknown Named Agents of the Federal Bureau of Narcotics* (1971). A Section 1983/Bivens type action allows any citizen who believes that he or she has been wrongfully treated to sue both state and federal officials (Chiabi, 1996). Such lawsuits most often involve violations of Fourth and Fourteenth Amendment guarantees against unreasonable search and seizure, false arrests, false imprisonment, assault and battery, invasion of privacy, and malicious prosecution (Kappeler, 1993). Civil liability has become a principal way to fight violations of constitutional rights, and Section 1983 has been found to be the most effective way to sue law enforcement officers, jail personnel, and other criminal justice practitioners.

Numerous lawsuits involving civil liability have been tried in court. The landmark decision is *Monell v. Department of Social Services* (1978), in which the Supreme Court ruled that both the government and the individual officer can be sued when the government is found responsible for a wrongdoing. In another landmark case, *Tennessee v. Garner* (1985), the U.S. Supreme Court held that deadly force can be used only if an officer is exposed to a situation in which he faces serious physical harm. In this case the Court introduced the *balancing test*. If a citizen's rights and interests outweigh the government's interests, there is likely to be liability for the use of physical force.

In *Pittman v. Nelms III* (1996), the Court held that immediate and serious danger has to be present before deadly force can be utilized. In *Berry v. Detroit* (1994), the Court held that 60 hours of initial firearms training, a certification examination and a written examination before graduation from the police academy, and additional deadly force policy manual, proper in-service training, and annual refresher courses after graduation were adequate training. However, some smaller police departments have been susceptible to claims for inadequate training simply because they lack resources. An example of this is *Davis v. Mason County* (1991), in which the county's training was found to be inadequate.

One of the most common forms of municipal liability is the allegation of inadequate or improper training of police officers. In searching for guidelines for devising an effective module for training liability standards, the case *City of Canton v. Harris* (1989) should be analyzed (Anon., 1990a). Police training was directly evaluated in this case. The plaintiff claimed that police had violated her constitutional rights by not giving her medical attention when she needed it and claimed that police commanders should have special training in order to make such medical determinations. The Supreme Court held that failure to train could be the basis of liability under Section 1983. The Supreme Court set the standard that " a municipality can be found liable under §1983 only when the municipality itself causes the constitutional violation at issue." Further: ". . . it may happen that in light of the duties assigned to specific officers or employees the need for more or different training is so obvious, and the inadequacy so likely to result in the violation of constitutional rights, that the policymakers of the city can reasonably be said to have been deliberately indifferent to the need. In that event, the failure to provide proper training may fairly be said to represent a policy for which the city is responsible, and for which the city may be held liable if it actually causes injury." Finally: "Moreover, for liability to attach in this circumstance, the identified deficiency in a city's training program must be closely related to the ultimate injury. Thus, in the case at hand, respondent must still prove that the deficiency in training actually caused the police officers' indifference to her medical needs."

Based on the case, the Court set forth some requisites for liability based on "deliberate indifference" (del Carmen and Kappeler, 1991):

1. The focus must be on the adequacy of the training program in relation to the tasks the particular officer must perform.
2. The fact that a particular officer may be trained unsatisfactorily will alone not result in city liability because the officer's shortcoming may have resulted from factors other than a faulty training program.
3. It is not sufficient to impose liability if it can be proved that an injury or accident could have been avoided if an officer had had better or more training.
4. The deficiency in a city's training program identified must be closely related to the ultimate injury.

The *Harris* case was used as a precedent when the U.S. First Circuit Court of Appeals decided the case of *Bordanaro v. McLeod* (1989). The case involved an allegation of police brutality during forcible entry to a motel room in an attempt to apprehend suspects previously involved in an altercation with an off-duty police officer. After the officers entered the room, the plaintiffs were beaten unconscious and one died from injuries.

The court found municipal liability failure to train based on the "deliberate indifference standard" as well as liability for the promotion of an "official" policy based on a custom of unconstitutional use of force and unlawful search and seizure. Liability was based on the following findings (del Carmen and Kappeler, 1991):

1. The department was operating under rules and regulations developed and distributed to the officers in the 1960s.
2. The department's rules and regulations failed to address modern issues in law enforcement.

3. The department failed to provide officers with training beyond that received in the police academy.

4. The city actively discouraged officers from seeking training.

5. There was no supervisory training.

6. The chief of police haphazardly meted out discipline and failed to discipline the officers in the current incident until after they were indicted.

7. There was no internal investigation of the incident until one year after its occurrence.

Training in the use of nondeadly weapons and techniques have also been evaluated several times by the Court. In *Kelly v. City of San Jose* (1988), the Court held that a police officer had received improper training in the use of a baton. Excessive use of a police dog was found in *Macleod v. Willie* (1993). Handcuffs were found to have been used excessively in *Cooper v. City of Virginia Beach* (1994), and in *Spann v. Rainey* (1993) the device misused was a flashlight.

Opponents of police civil liability have argued that being able to sue law practitioners can result in a police force that is afraid of acting. Most recently, in a case involving a number of sexual and physical attacks during a Puerto Rican parade in Central Park in New York City, the issue of police officers not responding to reports of misconduct has been raised as an example in which other cases, with much media coverage, made the officers present in the park afraid of involvement for fear of potential consequences and accusations of racism. On the other hand, proponents argue that even though litigation against police can be dangerous and costly (great resources are spent by the justice system to resolve these cases), there are several benefits. One argument states that when the government takes on the responsibility of providing service and protecting citizens, they can also be held accountable for any wrongdoing. Further, if citizens can be sued, so can law enforcement. Another argument points out that there must be consequences for police misconduct; if not, lawless police behavior can develop. Overall, proponents argue that allowing citizens to sue police enforcement practitioners through avenues such as Section 1983 has resulted in better and more effective police training and more responsible law enforcement practices.

Kappeler et al. (1993) conducted a content analysis of police civil liability cases handed down by U.S. federal district courts, between the years 1978 and 1990. Their research focused exclusively on published federal litigation brought against the police under Title 42, Section 1983 of the United States Code, as opposed to civil liability under state or federal tort law. Kappeler et al. point to the fact that some lawyers and citizens are quick to bring litigation against the government and its most visible agents, the police, which together with a rise in court findings of police liability has led to a perception that civil liability can be an occupational hazard even for well-meaning officers and professional departments.

Content analysis of the 1359 cases brought against police during the years 1978 to 1990 found that the police prevailed in 52 percent of the published Section 1983 cases decided by federal district courts. The success varied according to the type of agency being sued; plaintiffs were successful in 50 percent of the cases brought against municipal law enforcement defendants, 44 percent of cases brought against state police agencies, 40 percent of cases brought against federal law enforcement agencies, and 49 percent of cases brought against special police forces. The awards ranged from $100 to $ 1,650,000, with a mode of $5,000. The researchers concluded that there have been two major increases in the volume of police liability cases, and although police departments have a good record of defending themselves in civil suits, this record does not seem quite as good as reported previously.

Ross (2000) performed a content analysis of 1525 failure-to-train Section 1983 cases between 1989 and 1999. Case analysis indicates that failure to train and failure to supervise were combined as managerial liability issues in 54 percent of cases. Average attorney fees, depending on the category or topic of the litigation, ranged from $34,800 to $100,500. Ross's suggestions for police liability training include the following recommendations:

1. Police administrators should routinely conduct internal assessments of recurring tasks performed by officers and supervisors. Based on an analysis of incident reports, calls for service, citizen complaints, disciplinary actions, and recent changes in job requirements, a biannual training should be provided to all officers and supervisors.
2. Police administrators are encouraged to revise those policies and procedures that parallel training topics. The training should be "realistic and incident based."
3. Supervisory training should be instituted, before and after promotion, at least biannually.
4. All training should be documented and accurate training records should be maintained.
5. Training scenarios for the FTO program should be developed based on the relevant high-liability areas.

A study by Worrall (1998) discusses the influence of police officials in cases of litigation and highlights three principal aspects: minority recruitment, method of civilian review, and community-oriented policing. The main finding of this study is that the adoption of community-oriented policing (COP) has resulted in fewer lawsuits against the police (COP, education, and training are the only three aspects that have been related to reducing the risk for lawsuits against the police). However, the study is limited in its scope, and the overall perceptions of police officers regarding the threat of liability appear to be at least as valid as the findings of this study. Some contradictory views regarding the potential impact of COP on police liability are addressed later in the chapter.

Franklin (1993) lists a number of reasons why the number of police officers sued every year has grown steadily in recent decades despite the increase in both the length and quality of training. The following list should serve as a reminder and a warning for police trainers and educators that the post hoc approach to training should be replaced with a much more proactive method.

1. Our society as a whole is more litigious than in the past.
2. Removal of police officers from "beats" has removed the human element from the police.
3. Four distinct changes in our justice system have enabled the "have-nots" to voice injustices:
   - Claims against the police have been filed in federal courts based on the "feeling" of being more impartial and expeditious.
   - A growing number of students graduating with a law degree has increased competition for work among lawyers, and the introduction of a contingency fee allows clients with no money to pursue their cases.
   - The increase in information availability has helped lead to an increase in litigation.
   - The legal status of the police with regard to "sovereign immunity," which at one time was considered to be an absolute defense for police departments, has changed.

Garrison (1995) surveyed the attitudes of university, municipal, and state police toward the threat of civil liability. The method used was a survey distributed to three law enforcement agencies (50 police officers in total). To the statement "police officers should not be subject to civil suits by citizens" 42 percent of officers agreed and 52 percent disagreed. To the statement "civil suits against police officers are an impediment to effective law enforcement" 48 percent answered "true" and 50 percent answered "false." Sixty-two percent of officers believed that civil law suits have an effect on the behavior of police officers in the field. To the question "among the top 10 thoughts that go through my mind when I stop a car or confront someone is the threat of a civil suit as a result of the actions I take," over 90 percent of the officers answered "false."

The findings of Garrison's study should be taken into consideration when devising training modules related to police ethics and integrity. The survey found that (Garrison, 1995, p. 19):

1. Police officers generally do not think about being sued for liability on a daily basis.
2. Police officers see the threat of civil liability as a deterrent to police misconduct.
3. State police officers are least supportive of the normal citizen's option to sue the police for liability.
4. Municipal officers are most supportive.
5. Police officers in general are evenly divided on the issue of whether civil liability results in effective law enforcement.

Gallagher (1990) identified six main points to consider to avoid a police officer's exposure to liability and attain a high level of professionalism. He focuses on the supervisors, referring to them as the "extension of management" and "liability gatekeepers." Effective training of supervisors must (p. 41):

1. Be pre- and postpromotional
2. Occur at least biannually
3. Be contextual, with frequent use of relevant hypothetical incidents
4. Be agencywide
5. Include a component on performance evaluation or, more correctly, how to evaluate subordinates' performance on a more sophisticated level
6. Emphasize liability management

Vaughn (1994) researched police civil liability for abandoning citizens in high-crime areas or in other high-risk situations of victimization. The study suggested that liability occurs more often if the victims are children or if serious physical injury is involved. As a result, the author argues that to avoid liability suits, police guidance and training for police personnel must improve.

Another important area of liability can be found recently in *abandonment cases*. The Supreme Court cases that are used as a precedent for police liability in abandonment cases are *Martinez v. California* (1980) and *DeShaney v. Winnebago County Department of Social Services* (1989). In the first case a young girl was murdered by a parolee, and the parole board was sued. However, the court found that the girl's constitutional rights had not been violated because she had been victimized by the parolee solely. In the second case, a 4-year-old child was beaten badly by his father, and the county social service agency was sued because they knew about prior abuse. The court rejected the claim but acknowledged that in certain cases the state has the duty of care and protection for particular individuals.

There are three primary categories in which the police have been found liable for abandonment: abandonment of vehicle occupants, abandonment of children, and abandonment of assault victims (Vaughn, 1994). In contrast, there are four categories in which the police were not found liable for abandonment: qualified immunity, misrepresentation of crime seriousness, incompetent police response, and plaintiff assuming the risk of abandonment (Vaughn, 1994).

Vaughn also describes factors that the courts consider in assessing liability in Section 1983 abandonment claims. He divides them into factors including the plaintiff and factors including the police.

- The plaintiff
  - The plaintiff's age and mental capacity
  - The seriousness and nature of the plaintiff's injuries
  - Whether the plaintiff was in police custody or the functional equivalent of custody
  - Whether the plaintiff's actions contributed to his or her injuries
- The police
  - Whether the officer's actions or inactions created or enhanced the plaintiff's danger
  - Whether departmental policy contributed to the plaintiff's injuries

- Whether the officer possessed actual knowledge of an impending attack on the plaintiff
- Whether the officer possessed any special knowledge that could have prevented or reduced the plaintiff's injuries

The author concludes that these factors easily can be incorporated into policies and training programs.

Silver (1993) identifies the following points that might be considered as departmental indifference for constitutional rights:

- Contemporary standards of relevant law enforcement duties not being incorporated into departmental regulations
- No in-service training
- No command training for new supervisors
- No operational guidelines
- Haphazard and infrequent discipline
- Ignoring requests for training by officers

Thibault et al. (1998) compiled a basic checklist for police managers to implement in order to prevent tort actions (p. 324):

1. Do not allow untrained officers to perform any field police duties. An untrained officer may be defined as any officer who has not completed basic school and field training successfully.
2. Training academies need to keep accurate records, including:
   - Lesson plans
   - Attendance of all students
   - Instructor qualifications
   - Test scores
   - All handout materials given to students
   - All films and videos shown to students
   - Videotape of student during his or her training in skill-related courses
3. Official departmental policies should be reflected in training. The use of deadly force should be presented within the context of course instruction on deadly physical force.
4. Lesson plans, policies, and instructional techniques need to be reviewed and updated.

To conclude this brief review, a word of caution must be added about the frequently erroneous perception, held by police officers and police trainers, regarding the legal immunity issue, and more specifically and frequently, the qualified immunity concept. The term *qualified immunity*, refers to the fact that officials are immune from liability unless the law clearly proscribes the actions taken by them. A case example of this problem can be found in *Anderson v. Creighton*, (1987). In this particular case FBI agent Anderson conducted a warrantless entry and search in a bank robbery suspect's house and was sued for violating the Fourth Amendment rights of the house owner. "The Supreme Court found that the relevant question is whether a reasonable police officer could have believed Anderson's warrant-less search to be lawful, in light of established legal principles and the information known to Anderson and the other officers. If a reasonable officer could have concluded that the circumstances established probable cause and exigent circumstances—even though they did not—Anderson is immunized" (Anon., 1990b).

Similar standards of *reasonable officer* are used in other cases of police liability, most frequently in use-of-force cases. The misperception about the outcome of a given case in which a standard of a reasonable officer is used has to do with the fact that the final verdict of whether or nor the officer acted in a reasonable manner or as any other reasonable officer would, is determined by a judge or a jury which have a very different view of what constitutes reasonable behavior on the part of the officer. (See Chapter 7 about what appears to be reasonable and to whom in the case of 41 shots.) Therefore, police trainers should be extremely cautious when discussing and explaining the qualified immunity defense so as not to create a false sense of security in an officer's mind.

Familiarity Breeds Contempt

When the author starts her class on police–community relations with the afore-mentioned statement, the eyes of students open wide, some in disbelief, some in surprise, yet others, actually the majority, in protest. No, they seem to be saying, familiarity brings people closer together: They understand each other better, and understanding brings about a better relationship and cooperation—the desired outcomes of the community-oriented police philosophy. However, if for a moment we address some of the unpleasant roles that police officers have to play in real-life policing, removed from theoretical pondering, such as those during which an officer has to arrest somebody's child, spouse, or parent, and/or to use coercive force (within the legal constraints of that use) on citizens, and/or offer some deterrence to "wanna-be" lawbreakers of various sorts, such as the fact that during some civil disorder incident, the fact that one knows a particular officer does not help. To the contrary, to be coerced physically to do something against your will by somebody you know on a personal basis, to receive a traffic ticket that carries not only a fine of a couple of hundred dollars, but also contributes two to four points on your driver's license is not going to generate respect toward a given officer or the organization that he or she represents.

Community-oriented policing endorses closer personal contacts with community members. As Franklin (1993) pointed out, the isolation of officers from the public might have been one reason for the public to feel more compelled to file lawsuits. However, on the flip side, the same philosophy might also increase the number of lawsuits. The public is socialized by its police agency to the fact that police officers are here to serve, provide, and protect. The officers essentially present an image of a mighty friend, who is there to "do no harm." To a certain degree, the public might even "buy" this heavenly image, but only up to a point—the point of becoming too friendly with the beat officer. When one becomes too friendly, one also raises expectations about loyalty, dependability, and allegiance, expectations that are valid indeed when one deals with a real true friend. When unfulfilled, the same expectations breed contempt toward the person who failed to fulfill them. Of course, it is rather extreme to assume that any given citizen views or will view a police officer as a personal friend. However, the philosophy of community-oriented policing is "working" very hard on creating precisely this image. How else, then, can one explain the emphasis on cooperation, understanding, shared ownership, and the call to provide information, help, and assist? A friend who betrayed you is a target of contempt, a stranger who hurt you is a target of anger, but both are potential targets for litigation. Since the full-fledged introduction of community-oriented policing, it has been the author's belief that this philosophy will contribute to an increase rather than a decrease in cases of liability against police departments.

Kappeler (1997) acknowledges the possibility of reliance and contact between citizens and the police (the way it has been endorsed by community policing) as a possible contributing factor to render police liable if they do not take measures to ensure the safety of particular citizens. If this assumption is even remotely valid, it has some tremendous implications on the way that police officers should be trained to avoid, or at least minimize, various liability traps.

Community policing philosophy aside, it appears that the more civilized and sophisticated we become as a society in general and as individuals, the more we resent the tool that we have deposited in the hands of our protectors: namely, the right to legitimate use of coercive force against us. The numerous litigations and allegations related to police use of excessive force offer some validation to this hypothesis. If this hypothesis has some validity, the way that police officers are trained in the use of force has to be revisited as well. For example, the focus should shift slightly from a mere technical approach of how to close a pair of handcuffs on somebody's wrist to a much more extensive psychologically and physiologically based approach that would concentrate on both the physical and psychological reactions of the handcuffed person.

In one of her classes, the author uses her own handcuffs to illustrate the kind of impact that a simple act of closing handcuffs on somebody's wrist can generate in the controlled and safe experimental environment of a college classroom. During numerous experiments, student reactions varied from sweating, swelling, and uneasiness to more extreme demands "take it off me immediately—I cannot take it a minute longer!" If police trainers devoted more time—much more time—to the psychological and physiological reactions of both parties, police officers and citizens, it is quite possible that the number of allegations against police brutality and excessive use of force would have been reduced significantly.

Finally, this author would be remiss if she did not present another view about developing proactive training modules that will eliminate potential liability. Kappeler (1997) points to the danger inherent in the fact that courts throughout the United States continue to recognize the legitimacy of training and education for the policing profession, so that departments that fail to develop elaborate training courses will be deemed unreasonable and therefore prone to liability. For example, the U.S. Supreme Court has recently mandated police training in a number of areas, such as use of force, basic medical care, and pursuit driving. A given department that does not implement extensive training in these areas may find itself in a precarious situation.

To conclude, however, this potential predicament does not pose any dilemma to this author, as it is her profound belief that if one cannot change things in a progressive manner by articulating commonsense arguments, perhaps one can change them by pointing to the potential and very real threat. While presenting a paper during an annual meeting of the Academy of Criminal Justice Sciences on a model of academy training that includes mandatory inclusion of an FTO program, the author was accused of advocating dissent against the Constitution of the United States. It appears that the only sacrilege committed there was the author's desire to introduce extensive professional training modules, standardized for all police officers, which would of course take away the power of decision making from many local "Napoleons" who view the police force as their small private army instead of viewing policing as a true and independent profession and allow it finally to get the recognition it deserves. Of course, there is no bigger sacrilege than telling somebody that he or she might need to relinquish some of his or her power and control by subjecting their police officers to uniform standards of training, even if the standards have the potential to reduce police liability lawsuits.

One final note: What appears to be of prime importance is not the number of cases won or lost or the overall negative balance to the agency's budget, but the detrimental impact that each case, whether won or lost, has on an individual police officer's morale.

SUGGESTIONS FOR UNIFORM STANDARDS FOR POLICE LIABILITY TRAINING TO MINIMIZE NEGLIGENT TRAINING CLAIMS

A number of recommendations developed by the author and based on a review of literature, personal interviews, and work experience are presented for future consideration.

1. Revised police liability training must be included in the basic academy training. The revision should include the following steps:
   - Classes on psychological and physiological reactions to the use of force as experienced by both police officers and citizens
   - Classes on the psychological perception of the "audience" witnessing the use of force confrontations between police and citizens (a classic example of the impact of this perception is the shooting of a mentally disturbed person in New York City in 1999 by a number of NYPD officers)

2. Mandated roll-call training, at reasonable time intervals, but at least once a month, which will include the following topics:
   - Updates on recent outcomes of liability cases
   - Review of "acting under the color of law" situations
3. Mandated advanced supervisory/management training courses in minimizing civil liability through:
   - Regular updating of rules and regulations and operational procedures
   - Reviewing recent outcomes of liability cases and mandating relevant in-service-training based on the findings
   - Examining and evaluating available research findings in the areas relevant to police work: for example, the findings on use of OCC (pepper spray), or responding to a mentally challenged population
   - Validating training through a job-task analysis
   - Recruiting, assigning, and evaluating top-quality instructors
   - Evaluating, assessing, and documenting all the training efforts

SUMMARY

This chapter was written with two intentions in mind, to prevent officers from becoming victims of liability but also to prevent them from becoming victimizers. A theoretical discussion of dangers inherent in numerous liability cases directed against police officers and police organizations was followed by an overview of liability rulings and cases based on state and federal laws and regulations.

Despite the fact that the overall number of high-profile liability cases related in some way to inadequate police training is in itself not that overwhelming, the negative publicity following the cases seems to perpetuate and penetrate the ethos of the police profession. To minimize damage to the idea of policing in general to a given organization, specifically to an individual police officer, a number of recommendations are offered. Although the overall direction of the recommendations offered seem to point to one direction only, that of police officers and police organizations, the message carried in the chapter is twofold: to prevent officers from becoming the victims but also the victimizers.

REFERENCES

Anderson v. Creighton, 41 Cr.L. 3396 (1987).

Anonymous (1990a). *Canton v. Harris* Determines Standard for Training Liability Cases. *Police Chief*, June, pp. 37–38.

Anonymous (1990b). The qualified immunity: how to never lose a §1983 lawsuit. *Police Chief*, June, p. 39.

Berry v. Detroit, 25 F.3d 1342 (1994).

Bivens v. Six Unknown Named Agents of the Federal Bureau of Narcotics, 403 U.S. 388 (1971).

Bordanaro v. McLeod, 871 F.2d 1151 No. 88-1563, U.S. Court of Appeals for the First Circuit. (1989).

Chiabi, D. K. (1996). Police civil liability: an analysis of Section 1983 actions in the eastern and southern districts of New York. *American Journal of Criminal Justice, 21*, 83–104.

City of Canton v. Harris, 489 U.S. 378 (1989).

Cooper v. City of Virginia Beach, 21 F.3d 421 (1994).

Davis v. Mason County, 112 S.Ct. 275 (1991).

del Carmen, R. V., and Kappeler, V. E. (1991). Municipal and police agencies as defendants: liability for official policy and custom. *American Journal of Police, 10*(1), 1–17.

DeShaney v. Winnebago County Department of Social Services, 109 S.Ct. 998 (1989).

Franklin, C. J. (1993). *The Police Officer's Guide to Civil Liability*. Springfield, IL: Charles C Thomas.

Gallagher, G. P. (1990). The six-layered liability protection system for police. *Police Chief*, June, pp. 40–44.

Garrison, A. H. (1995). Law enforcement civil liability under federal law and attitudes on civil liability: a survey of university, municipal and state police officers. *Police Studies, 18*, 19–37.

Kappeler, V. E. (1993). *Critical Issues in Police Civil Liability*. Prospect Heights, IL: Waveland Press.

Kappeler, V. E. (1997). *Critical Issues in Police Civil Liability*, 2nd ed. Prospect Heights, IL: Waveland Press.

Kappeler V. E., Kappeler, S. F., and del Carmen, R. V. (1993). A content analysis of police civil liability cases: decisions of the federal district courts, 1978–1990. *Journal of Criminal Justice, 21*. 325–337.

Kelly v. City of San Jose, C-86-205260-RPA (1988).

Macleod v. Willie, 916710 AI, Circuit Court (1993).

Martinez v. California, 44 U.S. 277 (1980).

Monell v. Department of Social Services, 436 U.S. 658 (1978).

Pittman v. Nelms III, 87 F.3d 116 (4th Cir. 1996).

Ross, D. L. (2000) Emerging trends in police failure to train liability. *Policing, 23*(2), 169–193.

Silver, I. (1993). *Police Civil Liability*. New York: Mathew Bender.

Spann v. Rainey, 987 F.2d 1110 (1993).

Tennessee v. Garner, 471 U.S. 1 (1985).

Thibault, E., Lynch, L. M., and McBride, R. B. (1998). *Proactive Police Management*. Upper Saddle River, NJ: Prentice Hall.

Vaughn, M. S. (1994). Police civil liability for abandonment in high crime areas and other high risk situations. *Journal of Criminal Justice, 22*, 407–424.

Vaughn, M. S., and Coomes, L. S. (1995). Police civil liability under section 1983: when do police officers act under color of law? *Journal of Criminal Justice, 23*, 395–415.

Worrall, J. L. (1998). Administrative determinants of civil liability lawsuits against municipal police departments: an exploratory analysis. *Crime and Delinquency, 44*, 295–313.

Police Training in Other Countries
A Comparative Perspective

Introduction

Most police agencies in the United States provide between three and six months of entry-level training to newly hired police officers. Due to fiscal constraints, training is kept to a bare minimum; however, one might speculate that other concerns play an important role in both the content and length of basic training. It is possible that the average career span of a police officer in the United States is significantly shorter than in some of the countries mentioned in this chapter. The length of one's career may allow for more extensive entry-level training as well as broad training for each successive promotion. Nevertheless, there are a number of countries in which an officer can retire after 20 or 25 years of service and still be exposed to wide-ranging and advanced training. Local political considerations might provide yet another insight.

The information contained in this chapter may already be obsolete when this book is published. In many progressive countries, including the United States, training is an ongoing and constantly evolving and revised venture; therefore, recent information about training curricula is probably already being revised. In some academies the curricula are revised from one session to another. Despite the inclusion of this "dated" information, the goal of the chapter is to provide the reader with concepts and ideas with which police forces around the world are experimenting now. Therefore, if somebody familiar with a particular training concept in one country finds this information out of date, this does not invalidate the importance of the material compiled, as it is quite common for a department to experiment with one training concept, abandon it for another, only to return to the first one a year later. Hopefully, the medley of ideas presented below will inspire some police trainers in this country as it has inspired the author of this book.

GERMANY

The material in this section is based on an official publication of the Police College in Villingen–Schwenningen, *Policing Overseas* (1997). Commentary on the material is that of the author.

Germany is composed of 15 federal states, each with its own police force. Only three forces are organized as federal police: the Federal Border Police, the Federal Criminal Investigation Office, and the House Police of the federal parliament.

The information presented below is illustrative of police training in the state of Baden–Wuerttemberg. It has over 10 million inhabitants and is the third-largest federal state in Germany. The police force is composed of 26,720 employees, who are divided into 22,239 police officers and 4481 detectives.

The training program takes place in five different training centers. These training centers work closely with the individual police departments. The philosophy of training is based on the idea that police officers need to have certain personal skills that cannot be developed during training but are only supported by training. This means that the applicant should have the basic personal skills before he or she joins the police force.

Basic Training Program

The entire program lasts for 30 months. Following are the various steps of the training:

- Six months of basic training at the police school
- Six months of advanced training I at the police school
- Three months of practical work I at the police department
- Six months of advanced training II at the police school
- Three months of practical work II at the police department
- Six months of a final training program (exams included)

Based on the combination of field work and school, a good relationship is expected to be established between theoretical knowledge and later patrol duty. This dual system is considered to be the only sensible one, since the knowledge acquired in school can be used during work at the police department while the trainee is still in a learning environment.

At the end of the training program the trainee should be able to deal with the usual tasks in a professional manner; for example:

- Should have a convincing personality
- Should stick to the law
- Should apply the laws according to the interests of the people who are being helped and the situation they are in
- Should be able to settle disputes with the help of the law, with instinct, and with the use of technical equipment when necessary

The desired "readiness" of the German recruit appears to be far above U.S. standards, as most police academies, with very few recently instituted exceptions, require a passing grade of 70 for a recruit to graduate. When translated to a letter grade, 70 is an equivalent of a C–. Therefore, people who we entrust with responsibility to protect our lives are required to perform only on a C level.

In comparing this training module to training concepts in the United States, it can be observed that the Germans combine basic academy training with the FTO program (as advocated by the author in Chapters 4 and 5).

Training Program for the Higher Ranks

Prior to 1993, a police officer had to serve several years in the lower ranks in order to be promoted. Since 1993 it has been possible to start a police career directly in a higher rank. The applicant has to pass an exam and go through a training program that lasts 46 months.

The training consists of the following two levels:

- Ten months of basic training at the police school
- Thirty-six months at a college (including practical work)

By the end of the training program the trainee has to fulfill the same requirements as a trainee of the lower ranks. In addition to the above-mentioned requirements, the police officer expected to gain an extremely high social ability and communicative skills in addition to specialized knowledge. This is considered to be necessary in order to perform well in a middle management environment.

The following topics are covered in the program.

- General knowledge
    - German (writing, punctuation, and communication)
    - English/French (conversation classes, to be able to deal with foreigners)

- Social knowledge
    - Political knowledge (democracy, constitution, current political events)
    - Psychology (basic concepts)
    - Leadership (philosophy of policing)
    - Ethics (of the police job)
- Law
    - Criminalistics and tactics
- Practical subjects
    - Sport
    - Swimming and rescue
    - Self-defense
    - Use of weapons
    - Operating training
- Comprehensive guidelines
    - Criminal defense
    - Traffic accidents and control
    - Patrol duty
    - Self-study

Additionally, practical subjects that prepare trainees for dealing with people are offered as well. These classes assist the trainees in using the knowledge that they have already acquired. The trainees get full board in the academy, and a special free-time program is offered to give them some sensible possibility to spend their leisure hours.

What appears to be particularly interesting about the German approach to training, in addition to the length of the basic and advanced training, is the inclusion of certain topics of instruction and the allocation of time to these topics. For example, counting all four phases of training—basic, advanced 1 and 2, and final—the recruit spends an average of between 7.5 percent and 4 percent of training time studying "political knowledge" as opposed to 7.5 percent during basic training, 4 percent during advanced 1, and only 1.5 percent during advanced 2 (studying the use of weapons). The right "balance" between old and new skills seems to be approached in the most appropriate manner.

IRELAND

Information about Irish police training is adapted from Murphy (1996); a speech delivered by the then Deputy Commissioner of the Garda, Patrick J. Murphy; and from the official publications of the Guarda Siochana.

During the past 25 years, Irish society has undergone considerable change, which has resulted in a changed role for the Guarda Siochana. The Guarda Siochana has become a very complex, highly specialized, and centralized institution, making education and training reform necessary in order to prepare the personnel required to serve in the multitude of posts in this complex systems.

The recognition of three important concepts provided a framework for the new training:

1. Openness toward the community
2. Concordance with the realities of police practice
3. The promotion of individual police professionalism

These demanding roles required broad training as well as a range of knowledge and skills, including:

- A sufficient theoretical knowledge and understanding of Guarda powers and procedures and of as much of the law as is required for their immediate duties
- A level of competence in the skills needed to apply their theoretical knowledge in the operational field
- A level of competence in the exercise of interpersonal interaction and communication skills
- A developed capacity to make prudent judgments in the exercise of their very extensive discretionary powers
- A level of competence in the effective utilization of the range of technical equipment that is on issue to the force
- Achievement and maintenance of a satisfactory level of physical fitness and competency in the exercise of the physical skills required to perform their duties efficiently
- An appreciation of the need for, and commitment to, continuing to extend the range of their knowledge and skills, thereby enhancing their personal and professional development and their contribution to the force and to the society they have chosen to serve
- An adequate awareness and appreciation of the broader historical, psychological, social, economic and political contexts in which the force operates
- An acknowledgment and appreciation of their social role in the community, including familiarity with the nature and location of the various voluntary and statutory social agencies that are willing to work with the Guarda Siochana

The training, which integrates theory and practice, lasts for 104 weeks and is organized into the following five phases:

Phase I. Twenty-two weeks of training is delivered at the Guarda Siochana College and is basically a foundation course where students receive an introduction to the core subjects in the curriculum: legal studies, policing studies, physical education studies, Gaelic, communications studies, social studies, technical studies, and European language studies (a choice of French or German).

Phase II. Consists of twenty-two weeks of practical training, during which the candidates are attached to selected Guarda stations throughout the country. In this phase, as in phase IV, the education/training program is supervised by training sergeants and assisting training sergeants, who are responsible for coordination and the weekly classroom instruction of students in the in-service schools. Phase II is divided into three subphases.

1. *12 weeks:* The trainees spend the first 12 weeks attached to a unit at their station. Each student is assigned a tutor Guarda who is an experienced and specially trained member and is responsible for guiding the student and introducing him or her to all aspects of operational Guarda work.
2. *10 weeks:* During the second half of this phase, each student is detailed for a 10-week period for various units and sections within the service, including divisional and district offices, the drug squad, crime investigation units, and juvenile diversion program offices.
3. *2 weeks:* During the last stage of phase II, each student spends two weeks on social placement with an external voluntary or statutory social agency. These may include the ambulance service, women's refuges, social service centers, residential units for mentally handicapped adults and children, Boy Scouts, Girls Guides, the Irish society for the prevention of cruelty to animals, addiction treatment centers, and various day and drop-in centers for the elderly and young people.

Phase III. During phase III the students are back at the Guarda Siochana College for 12 weeks, where the aim is to enable them to gain a deeper understanding of the theories and concepts studied during phase I by applying them to practical situations that they encountered in phase II. Upon successful completion of phase III, the students are attested to An Guarda Siochana at a ceremony conducted at the college and become probationary Gardai.

Phase IV. During this probationary stage, probationer Gardai are attached to stations as full members of the service. Probationers spend 36 weeks engaging in as wide a range of duties as possible in order to broaden their professional experience. Close supervision is given during this phase, and they also attend lectures for two days a month at their local in-service schools.

Phase V. The aim of this six-week phase at the Guarda Siochana College is to allow probationers, through experiential learning methods, to share their experience and further develop their knowledge of policing. Toward the end of this phase they sit for their final examinations. Upon successful completion of these examinations, the probationers participate in a graduation ceremony to mark the end of their two-year formal program.

Probably the most innovative theme in the Irish training is phase II, the practical training, during which the probationers spend two weeks on a social placement. It is hard to think of a more imaginative way to expose the officers to the "softer" side of their work, one so critically important in the development of multicultural and community policing orientation.

Turkey

Information about Turkish Police training was provided by Ibrahim Cerrah (2000) of the National Police Academy (personal communication, 2000). The Turkish Police is a national police force, with centralized police training. There are two types of police training institutions in Turkey: the National Police Academy and Police Schools (Police Academies).

Police Schools

Turkish Police Schools are the equivalent of police academies in the United States. There are 22 to 24 separate police schools in Turkey located in different parts of the country. Although they are locally based, they are all controlled by the Department of Training at the General Directorate of Turkish Police Headquarters, in Ankara. Cadets admitted to the police schools are selected among high school graduates. The applicants have to pass the following four stages. The first stage is the background check, followed by an interview, a written exam, and finally, a physical test and/or exam.

The length of police training in Turkey is one full academic year, which is about a nine-month period. Within this period cadets receive theoretical training in such areas of police practice as:

- Introduction to law
- Penal law, criminal law, and procedures
- Human rights
- Firearms
- Police tactics and techniques
- Intelligence gathering
- Physical training, self-defense (martial arts), crowd behavior, etc.

Cadets who attend the nine months of training and pass all the exams graduate as ordinary police officers with no rank or prospect of promotion. They are appointed to any part of the country and to any police department or unit.

They receive on-the-job training very rarely. If an officer completes a university education while he or she is serving, he or she gets the chance of promotion to a higher rank. Occasionally, those who complete university education courses are offered for promotion to a higher rank. Following attendance at these courses, like any other police academy graduates, they have a chance of receiving promotion every three or four years.

National Police Academy

The Turkish Police Academy is the equivalent of the Police Staff College in the United Kingdom or police universities in some countries. There is only one police academy located in Ankara. The head of the police academy is called the president, a rank second only to that of the general director of the Turkish police. Cadets admitted to the police academy come from two sources. The first is the police college. In Ankara there is only one high school–level police boarding school, the Ankara Police College. The school admits students with a secondary school background. It gives four years of high school education without any police training. The differences between the police college and any other high school are that students at the police college have to wear uniforms, and it is a boarding school. When students graduate from the police college, almost all of them are admitted to the National Police Academy. If some of the graduates of the police college decide to attend a civilian university, they can do so. Almost 80 percent of the police academy cadets come from the Ankara police college. The remaining 20 percent of the cadets come from ordinary high school graduates.

The second source for police academy cadets are high school graduates. Those who took a national university exam and get the required grades are entitled to apply to the police acadmy. The grade required to apply to the police academy is relatively high and changes slightly every year. Applicants with an ordinary high school background also have to pass the following four stages; background check, interview, written exam, and finally, a physical test or/and exam. The length of training at the police academy is four full academic years. Within this period, similar to that of the police schools, cadets receive theoretical training in such areas of police practices as:

- Introduction to law, penal law, criminal law, and procedures
- Human rights
- Weapons training
- Police tactics and techniques
- Intelligence gathering
- Physical training, self-defense (martial arts), and crowd behavior (similar to the classes offered by police schools)

In addition to the classes given at police schools, other classes are offered, such as foreign languages (English, French, German), sociology, social psychology, economics, statistic, computer training, criminalistics, and criminology. Cadets who completed four years of training and passed all the exams made within this period graduate as police sergeant. They are appointed to any part of the country and to any police department or unit. Like ordinary police officers, they receive on-the-job training very rarely.

Every three or four years they receive automatic promotion up to the rank of police chief. For the aforementioned reason, the Turkish police presently has more police chiefs than needed. The numbers are getting increasingly higher every year. This is one of the problems that the Turkish police is facing and needs to solve urgently. As a temporary solution, some of the police officers who have police chief ranks are appointed to a department that serves as a reservoir (Cerrah, 2000).

The length of training required from police officers in Turkey, on both the basic and advanced levels, appears to be extremely impressive, especially when compared to the length of training in the United States. The four years of training at the police academy can definitely serve as a model to be adopted for management/executive police training in lieu of various forms of management training that police agencies in the United States tend to experiment with. The Turkish police training has been revised in 2001. Major changes will be introduced at the end of 2001.

THE NETHERLANDS

Information about Dutch police training was adapted from an official Dutch police publication about the state of policing in the Netherlands (Dutch Ministry of Justice, 1999).

In recent years the Dutch police service has undergone radical changes. The majority of these have resulted from reorganization of the service: The municipal and national police forces have been reformed into a single organization comprising 26 forces, 25 regional forces and a national police service agency. The national force provides certain support services to the 25 regional forces. The size of a regional force depends on factors such as population size, crime levels, and building density. Each regional force breaks down into a number of district divisions, each headed by a district commander with a small staff. Districts are usually subdivided into basic units.

Prior to joining the police service, all recruits follow one of three basic courses of training: (1) training as a surveillance officer, (2) training as a constable, and (3) training as a senior police officer. There are also a number of additional courses provided by several institutes. The responsibility for the implementation and quality of training lies with the LSOP, the National Police Selection and Training Institute. The ministers under whose auspices the police fall determine the exit qualifications and the length of training. Examinations are supervised by inspectors, which thus ensures that the qualifications are actually met.

The LSOP was set up in 1992 to oversee and coordinate the various police training courses. Its actual work extends much further, however. The institutes provide public information on working for the police service and assists regional forces in recruitment and selection. It also advises on organizational and educational matters and does research on police training. The LSOP is run by a board comprised of representatives from the police, government, the public prosecutions department, trade unions, and industry. The individual training institutes also have supervisory committees, which are composed in a similar manner.

The length and exit qualifications of the various courses of training for the police are determined by the Ministries of Justice and of the Interior. Quality controls are also built in, as the examinations must by law be supervised by inspectors. There are sufficient guarantees that the desired final levels are actually reached.

Training as a Surveillance Officer. Training for the rank of surveillance officer is a new form of basic training for officers whose duties will lie primarily in the field of monitoring and prevention. It lasts six months and is comprised of 6 modules. The LSOP can, if required, design supplementary optional modules for specific functions. Training can be extended through optional modules and periods of placement to a maximum of one year, and always culminates with an examination. Surveillance officers have the competence to investigate offenses.

Training as a Constable. This is also a new form of basic training for police officers, still being updated. It has been designed to reflect changes in the professional profile of police officers and will probably last 16 months. Central course material can be supplemented with material specific to a certain region. On-the-job training is being extended at the specific request of the forces. Trainees experience several periods of on-the-job training, thus reducing the gap between theory and practice. An advantage of this new form of training is that it will be easier for surveillance officers to become constables, as they may be granted exemption from the three first modules of the course.

Both of the training courses described above are offered at the five police academies. The Ministries of Interior and of Justice pay for the training. These academies also provide some follow-up courses and educational services for other parties.

Training as a Senior Police Officer. This basic training course is designed to prepare senior police officers for management positions within the service. The course is offered by the Netherlands Police Academy, an educational and research institute located in Apeldroorn. There are four types of courses and study routes to chose from. A course takes between two and four months, interspersed with periods of on-the-job training with the candidate's own force. Candidates are usually experienced officers with several years of experience who already head a district team or a force unit and are sent by their forces to undergo advanced training.

In addition to the centers providing basic training, there are training institutes that provide other courses. One such institute is the Police Study Center in Warnsveld, which organizes conferences, courses, and seminars for senior police officers and administrators. It provides a backup in reorganization processes and advises on matters relating to finance, information, personnel, equal opportunities, and culture. Individual consultation is also possible.

The Junior Officer Training School and the central Training Institute provide practical courses aimed primarily at the middle ranks of the police. The police academy is in the process of merging with these institutes to form a new Netherlands Police Academy. Another initiative is also on its way, to devise training for the higher ranks in an attempt to modernize the training of senior managers within the Dutch police.

Police Institute for Public Order and Security.   Many regional forces assign officers who have completed their basic training to specialized courses, to qualify them to serve in mobile units. These courses, both basic and advanced, are offered at the Police Institute for Public Order and Security. The institute has its own custom-built "village" in which almost any circumstances can be simulated for the purpose of practical training and a fully equipped shooting range where recruits can practice shooting from moving vehicles.

Police Traffic Institute.   The Police Traffic Institute offers numerous courses, ranging from elementary to advanced, on all aspects of road safety, traffic, and driving techniques. It provides driving and surveillance courses relating to cars, motorcycles, and special vehicles. The institute also trains arrest teams and security services. The staff provides expert information and advice on vehicle technology, traffic legislation, and traffic management. It has its own up-to-date test circuit and a skid track.

Criminal Investigation School.   The school offers courses on criminal investigation techniques, develops the training curriculum, and advises on crime prevention. The school also produces teaching materials and organizes conferences. The police service is its biggest customer, but its services are also utilized by the Public Prosecutions Department and the Ministry of Justice.

The 40 or more courses offered are aimed both at officers who perform day-to-day police tasks and those in managerial positions. The institute also trains instructors to teach its courses. The courses range from basic criminal investigation to highly specialized courses in such fields as fraud, drugs, and technical investigation (Dutch Ministry of Justice, 1999).

What appears to be the most important trend to be considered from the Dutch training is their approach to specialized training. Bearing in mind Vollmer's quote, and addressing the "desired" skills of a "good police officer," there is no need for all the officers to be specialists in every field. The separately designed training modules for surveillance and constables seem to target the problem of overspecialization, which essentially becomes underspecialization for most officers in the United States. Again, as in other countries, the length of time devoted to advanced training is not comparable to training in the United States.

FINLAND

Information about Finnish training is adapted from two official Finnish police publications about the state of police training in Finland (Finnish Police, 1999).

The police force in Finland is a national police force headed by the National Police Commissioner. There are five provincial administration units of the police, 90 local police departments, the police district of the Aland Islands, and the police department of Helsinki, the capital city of Finland. Police training is delivered in police school and the police college of Finland.

Police School

The police school in Finland is an institute subordinated to the Ministry of Interior. It is responsible for student selection for the basic degree in police studies, providing basic training and the more advanced noncommissioned officer's degree. It also provides specialized professional training in the police field.

Noncommissioned Officer's Degree. The scope of the noncommissioned officer's degree is a minimum of 20 credits. The study program consists of mandatory courses and electives. Holders of the degree are qualified to apply for the posts of sergeants, detective sergeants, and senior detectives with no supervisory tasks. The training program of each student aiming at the noncommissioned officer's degree is part of a personal training plan prepared on the basis of interviews conducted with the student's superior. An admission criterion for the training program is that a student is a holder of the basic degree in police studies and she or he has work experience of at least two years in police duties. After completing a noncommissioned officer's degree, a police officer may apply for the post of sergeant.

Basic Degree in the Police Studies. The studies incorporate a total of 82 credits divided into three modules: (1) Basic Studies 1, (2) Fieldwork, and (3) Basic Studies 2. The fieldwork takes place in a local district police department assigned by the police school. A written essay of five credits is included as a requirement for the basic degree. The total duration of the studies is two years. Students may live in the student residence. Those attending Basic Studies 1 and 2 have meals and medical care arranged by the police school. Police cadets attending Basic Studies 1 receive a daily allowance confirmed by the Ministry of Interior. Those attending Basic Studies 2 receive a salary confirmed by the Ministry of Interior.

The structure of the basic degree in police studies is as follows:

- Basic Studies 1 (42 credits, one year)
  - *Period 1:* Society and the Police (six credits)
  - *Period 2:* Basics of Police Activities (six credits)
  - *Period 3:* Field Activities and Emergencies (six credits)
  - *Period 4:* Customer Service and Dispatch Activities (six credits)

- *Period 5:* Basics in Crime Investigation (six credits)
- *Period 6:* Pretrial Investigation (six credits)
- *Period 7:* Control and Preventive Measures (six credits)
- Fieldwork (17 credits, six months)
- Basic Studies 2 (18 credits, six months)
  - *Period 8:* Field Activities (six credits)
  - *Period 9:* Crime Prevention and Other Preventive Measures (six credits)
  - *Period 10:* Crime Investigation (six credits)

After completing basic training, a cadet graduating from the police school is entitled to apply for the post of senior constable.

Further Degrees in Police Studies.   There are two further degrees in police studies: a commanding officer's degree and a master's degree. A commanding officer's degree is a college degree. Completing studies for the degree takes three years. After graduation, a police officer holding a degree may apply for the post of inspector or superintendent. However, applicants for the post of police chief are required to have a second university degree, a master's degree. Those who have taken a commanding officer's degree may apply to universities to continue their studies with the aim of obtaining a master's degree in administrative sciences.

The curriculum for police chief offered by the Police College of Finland consists of complementary training designed to fit the personal needs of students. This training is aimed specifically at those who have already taken a master's degree in legal sciences and who apply for the post of police chief without actual training in police studies.

Police College of Finland

The Police College of Finland is a college with a vocational focus. The curriculum and degree requirements have been set up to correspond to a college degree. Graduates of the police college have an opportunity to continue their studies at a university for higher degrees, up to a doctor's degree. The college focuses on providing training for commanding officers and for the police administration in general, with special emphasis on commanding-level police officers and police chiefs. It offers both scientific and practical courses in subjects related to the police field. In addition, it carries out research projects and has close contact with other colleges and universities; together, these two factors establish the basis for training in police studies.

The highest-level police studies and related research are conducted in close cooperation with universities. When a student is taking a commanding officer's degree at the police college, the university education is provided by the universities of Tampere and Turku. The police college has contracts concerning scientific education in the field of police studies with the two universities. Two professorships have been established for training and research in the field. In addition to these two universities, there are respective projects conducted with other universities as well.

The reform in police studies in Finland made it possible for persons having no prior police training to apply for advanced police training. Qualifications for a post of police chief have been amended in a way that an applicant must hold either: (1) a master's degree in legal sciences or (2) a commanding officer's degree and a second degree in some other suitable field. This amendment enables future recruits for the police service to have a more versatile background. The Finish practice in this respect corresponds to the one in use in many other police forces, such as Germany, the Netherlands, and Great Britain.

Commanding Officer's Degree.   Students are selected on the basis of an entrance test and an aptitude test. A noncommissioned officer's degree and a minimum of two years of work experience are the requirements for taking the entrance test. Applicants who pass the entrance test are then assigned to paid training for the period of their studies.

A commanding officer's degree completed at the police college is a college degree. The requirement for the degree is 120 credits, so it is possible to complete the degree in three years. According to the Finnish education system, 1 credit equals 40 working hours. Studying takes place mainly in periods of intensive training at the facilities of the police college. However, a significant part of the curriculum is conducted independently as supervised distance learning.

The curriculum is divided in the following manner:

1. Administrative sciences   23 percent
2. Legal studies   21 percent
3. Management of police activities   15 percent
4. General studies   14 percent
5. Sociology of law   11 percent
6. General management   10 percent
7. Methodology   6 percent

Study Program for Police Chiefs and Complementary Training.   A significant part of the duties assigned to the Police College of Finland is connected to training provided for police chiefs and complementary training provided for the police organization. The curriculum for police chiefs contains systematic orientation studies and advanced complementary training in subjects related to professional tasks. For those who have not been trained in the police school, studies in police administration and management of police activities are emphasized.

Studying is prescheduled and takes place after a student has been appointed to a vacant post. Depending on a person's previous education and working experience, the studies take from 30 to 98 credits. Those who have a commanding officer's degree have the opportunity to continue their studies at universities, particularly at the universities of Turku and Tampere. Studies at the Police College of Finland receive full credit. By completing a unit of a minimum of 40 credits, a student will have the 160 credits required for the master's degree.

The approach of the Finish police to police training is worth a few words of praise and certainly brings it much closer to the desired state. The fact that the training is treated basically as a college degree instills in recruits from the very start the right attitude for the profession. For most police recruits in the United States, a distinction between academy and college training is quite obvious. In Finland, a graduate of the police academy has 82 credits and basically, carries a degree. A difference between a degree with 82 credits and certification from a police academy can indeed make a difference in the way that both police and the public view this profession.

POLAND

Information about Polish police training was adapted from an article written by the chief of the Polish police, General Jan Michna (1999), published in the *Criminal Justice International* and compiled by the author during her visits to the training centers in Poland in 1998 through 2000 (Haberfeld, 1999, 2000).

The Polish Police Force is a national force, comprised of about 100,000 sworn officers. After 1990, the Polish Police Service began to undergo an intensive process of staff changes. In the period 1990 to 1996, 46 percent of the staff was replaced. A need for effective, fast, and methodologically sound training of the new police staff became paramount to the force's successful transformation from the communist militia (until 1989) to a newly established police (1990) based on democratic principles of government.

Since January 1999, the Polish police service has been operating under a new structure, reflecting the responsibilities and hierarchical relations provided for under the relevant statutes passed by the Sejm (the Polish parliament). The police is composed of three primary types of forces: criminal, preventive, and organizational (logistics and technical support). Under the general headquarters of Police there are 16 district headquarters, 1 metropolitan police headquarters, 263 county headquarters, 65 municipal headquarters, and 1793 police stations, whose numbers change as some are liquidated or consolidated.

At present, professional training of police officers is conducted at the following levels:

- *Basic training:* aimed at introducing all new police officers to the basic requirements of the job; conducted in 12 police centers
- *Specialist training:* highly specialized training aimed at preparing police specialists to carry out functions in various police forces

- *Training for noncommissioned officers:* aimed at developing professional qualifications and serving as a requirement for promotion to the rank of noncommissioned officer
- *Training for candidates:* designed to ensure a further development of professional qualifications of outstanding police officers, and enabling promotion
- *Training for officers:* designed for police officers with an excellent record, several years of service, and completed training at all the lower levels of training

Programs of police officer training in Poland are of modular character. Training at a police school or center is interspersed with on-the-job training in police units, where trainees can test their knowledge and practice skills learned in the course of training. Among the commonly used participation teaching methods are interactive lectures, simulations, and role playing.

Training is conducted in the following institutions:

1. The Police Training School in Slopsk trains police officers in prevention at the level of noncommissioned officer and candidate and offers development training for managerial staff up to the level of head of prevention department at county headquarters and police chiefs of category III police stations.

2. The Police Training School in Pioa trains police officers in criminal investigation at the level of noncommissioned officer and candidate and offers development training for managerial staff up to the level of head of criminal investigation department at county headquarters.

3. The Police Training Center in Legionowo trains police officers in traffic regulation, criminal techniques, and observation at the level of noncommissioned officer and candidate. Additionally, it offers basic training for police officers with tertiary education and specialist courses in:
   - Criminal investigation
   - Juvenile delinquency
   - Intervention techniques
   - Water police
   - Mine disposal and pyrotechnics
   - Patrolling and intervention
   - Criminal prevention
   - Information technology

   In July 1998 the International Center for Specialist Police Training was opened at the Police Training Center in Legionowo. The center offers development courses for police officers from countries of central and eastern Europe specializing in fighting the most serious crimes, including cross-border crimes. The center also serves as a meeting and training place for representatives of other bodies for the protection of public order and law enforcement bodies from European countries.

4. The Police Training School in Katowice trains police officers of the preventive force for large urban areas at three levels: basic, noncommissioned officer, and candidate.

5. The Higher Police Training School in Szczytno trains police officers at the officer level for the purpose of basic police forces and prepares them for specialist posts and lower-level managerial posts. Specialist training is conducted by the Institute of Prevention and the Institute of Criminal Investigation. General subjects, such as law, introduction to philosophy, economics, sociology, psychology, professional ethics, police statistics, and foreign languages, are taught by the Institute of Law and Social Sciences and the Institute for the Education of the Police Managerial Staff.

Professional development is an important element in the process of police staff training in Poland. It is carried out mainly by participation in specialist courses, which deepen professional expertise and provide an opportunity to learn about the latest methods and forms of crime fighting.

Professional development is also pursued by organizing national competitions, tournaments, and professional and fitness contests. The best police officers of a given specialty in provincial events are qualified for the finals.

What is especially impressive about the way training is delivered to the Polish police officers is the approach to specialist and advanced training. Both specialist/ specialized and advanced training are offered by the police training centers and not by private or semiprivate entities (with no or very little accountability). which offer this type of training in the United States. The importance of having specialized and advanced training delivered by a centralized training center (similar to the system in other countries) cannot be overemphasized, and its importance was addressed in earlier chapters.

CANADA

When the Canadian police forces moved into the twenty-first century, community policing was already firmly established as the dominant policing orientation. The preponderance of Canada's 420 police agencies have either officially endorsed or otherwise supported this concept as the most appropriate approach for contemporary policing, and the endorsement of the idea can be observed in their methods of training.

Canada's five largest police departments are (Rosenbaum, 1995):

1. Royal Canadian Mounted Police
2. Ontario Provincial Police
3. Sûreté de Québec

4. Metro Toronto Police (recently renamed Toronto Police Services; Haberfeld, 1999a)

5. Montreal Police

Police work in Canada has traditionally been carried out at three levels: federal, provincial/territorial, and municipal. In recent years another level has been added, policing by aboriginal police forces on First Nations lands. The total number of police officers in Canada was 54,311 in 1996.

Recruit Training Models

The following description of Canadian training is adapted from Griffiths et al. (1999)

There are currently four models of basic training for police recruits operating in Canada.

Model 1: Separation of Police Education and Training from Mainstream Adult Education. Recruits and police officers who have at least a grade 12 level of education are trained in institutions that are separate and independent. These include the Royal Canadian Mounted Police (RCMP) Training Academy in Regina, the Ontario Police College in Aylmer, and police training academies in major cities where there are no provincial training institutes. Typical recruit training in the Ontario Police College in Aylmer consists of the following six levels:

- *Level I:* one to two weeks of field training by individual police agencies
- *Level II:* 60 days of classroom teaching
- *Level III:* three months of field training
- *Level IV:* 14 days of classroom training
- *Level V:* general duties
- *Level VI:* optional specialized training at the college and three years of service-established eligibility for promotion to the rank of first-class constable

RCMP Cadet Training Program. The training depot in Regina offers a competency-based training program centered on the CAPRA model, which identifies basic organizational values.

- C (client)
  - Ethics and professionalism/integrity
  - Dress, cleanliness, and deportment
  - Client orientation/service
  - Problem definition
  - Communication skills

- A (acquiring and analyzing information)
  - Knowledge of law, policy, and procedures
  - Information gathering
  - Records management
  - Crime scene investigation and evidence gathering
  - Conduct of searches
- P (partnership)
  - Team building and facilitation
  - Consultation, negotiation, and conflict resolution
  - Interagency and multidisciplinary cooperation
  - Planning and coordination
- R (response)
  - Risk management
  - Public and police safety
  - Decision making
  - Care and handling, arrest and release of suspects and prisoners
  - Testimony in court
  - Crime prevention/alternative to enforcement strategies
  - Victim relations and services

Skills:

- Tactical maneuvers and operations
  - Driving
  - Firearms
  - Police defensive tactics
  - Fitness and lifestyle
  - Lifesaving
- A (assessment for continuous improvement)
  - Monitoring and contingency planning
  - Incident review and self-evaluation

The cadet training program consists of 749 hours of training broken down as follows:

Applied police science	409 hours
Police defensive tactics	72 hours
Fitness	75 hours
Firearms	69 hours
Police driving	70 hours
Drill, deportment, and tactical exercises	54 hours

Model 2: Model 1 Delivered on a University Campus.   In this model, police staff teach classes on police administration and procedures, academic staff teach classes on criminal justice and social services, and lawyers and judges teach classes on criminal law, evidence, and procedure.

Model 3: Holistic Approach to Recruit Training.   This model is based on the assumption that police recruits should be exposed to the entire criminal justice system rather than just to the field of policing. An example of this model is the Justice Institute in British Columbia, which provides training for police officers, probation officers, court staff, and correction personnel. The basis for the training block is altering classroom teaching with field experience. Recruit training in the Justice Institute of British Columbia is composed of a five-block, three-year training program.

- *Block I:* 12 weeks at the academy; basic police training, legal studies, traffic, driving, firearms, community relations, investigation and patrol techniques, arrest and control, physical training, foot drill, dress and deportment
- *Block II:* nine weeks of field training; in the recruit's police service, under the supervision of a field trainer, usually consisting of a working two-person unit
- *Block III:* nine weeks at the academy; simulations and techniques concentrating on people skills and community relations

Successful candidates are qualified municipal constables and are able to work alone upon return to their departments.

- *Block IV:* two weeks at the academy; requalification, review of previous training, updates, investigative techniques
- *Block V:* one week at the academy; requalification, self-paced study package, legal/traffic updates (exam)

Successful candidates are Certified Municipal Constables (first class constables).

Model 4: The Quebec Model.   Police training in Quebec attempts to integrate police education with adult education. New recruits are required to complete a three-year college program and obtain a diploma of collegial studies, which includes general academic courses and instruction in criminology, policing, and law. Following this, candidates are send to the Quebec Police Institute in Nicolet.

The training center places a strong emphasis on hard police skills, including driver training, firearms, and arrest and control techniques, as well as soft skills such as physical fitness and community relations. The intent of the training center is to break down knowledge into a variety of disciplines that can be enhanced when the recruit begins his or her field training on the street. The recruits in the Nicolet Institute experience the work of police officers in the field in a simulated station located on the premises. The training program at Nicolet incorporates a competency-based learning approach.

Regardless of the specific training center, police officers in Canada generally receive instruction in the following areas:

- Legal studies
- Investigation and patrol
- Community relations
- Firearms training
- Traffic studies
- Driver training
- Physical training (including use of force)

Ontario Police Foundation Program

In 1998, the province of Ontario implemented an innovative approach to police training centered on a two-year foundation's program offered primarily at community colleges. The Foundations Program, which is supported by the Ontario Association of Chiefs of Police, was made a mandatory program for all police applicants in 2001. The first year of the program includes courses in general areas such as criminology and psychology, while the second-year courses focus more on specific areas of police work, including criminal law, case investigation, and conflict resolution.

Upon successful completion of the two-year course, students will receive a foundations training certificate and then write a provincial examination. Applicants who pass the examination are eligible for employment by any police service in Ontario. Successful applicants to police services will then be sent for further training at the Ontario Police College.

Operational Field Training.   Upon completion of recruit training, individual officers are usually assigned to general patrol duties for a period of roughly three to five years. After this initial period the officer receives additional in-service training. For the most part, in-service training is designed for learning the basic skills of police work by complementing recruit training and the hands-on experience received in the field.

During this second component of the training/learning process, known as *operational field training*, the recruit learns to apply the basic principles taught at the training center. Under the guidance of a specially trained senior officer, the recruit is exposed to a variety of general police work. A primary objective of the field trainer is to enhance the skills and knowledge that the recruit has learned at the academy. Despite the fact that police trainers play a significant role in the training process and influence the attitude and style of policing developed by the new recruit, selection of the trainers is often haphazard.

In-Service Training.   Training provided to police officers during the course of their career is usually conducted by individual police agencies or by provincial training centers. A key issue surrounding this type of training is whether it should be mandatory or optional. Some police services mandate a specific number of training hours to be completed; others permit the police officer to do so if they desire. The trend in Canadian police services is toward integrating in-service training with career development and requiring that officers achieve certain educational and training competencies in order to apply for promotion and advancement. A variety of in-service specialty courses are available to Canadian police officers, including:

- Administrative skills
  - Effective writing for supervisors
  - Selection and interviewing
  - Media relations
  - Supervisory training
  - Effective presentations
  - Field commander training
  - Firearms range officer
- General patrol skills
  - Bicycle patrol
  - Firearms
  - Effective writing
  - Field training for recruits
  - School liaison
- Investigation-related skills
  - General police investigations
  - Child abuse investigations
  - Crisis (hostage negotiations)
  - Statement analysis
  - Surveillance
  - Fraud investigation
  - Interviews and interrogation
  - Drug enforcement
- Traffic-related skills
  - Breathalyzer data masters
  - Approved screening device calibrators
  - Accident investigation (level II)
  - Accident investigation (level III)
  - Motorcycle operation
  - Heavy truck enforcement

Although there are many opportunities for police officers to participate in in-service training and continuing education programs, few police services have the capacity to restrain officers or to conduct reassessments of basic policing skills after officers have been on the job for several years.

Canadian Police College

The Canadian Police College, located in Ottawa, is operated by the federal government under the federal ministry of the solicitor general. The education and training programs offered are national in scope and are designed to provide police officers from municipal and provincial police services, as well as RCMP members, with upgrading and development programs, research, information, and advice. The specialized training courses are typically two to three weeks in duration and offer the following topics:

- Advanced collision analysis
- Advanced vehicle theft investigative techniques
- Automated fingerprint classification
- Clandestine laboratories investigation
- Collision reconstruction
- Commanders: hostage/barricaded persons
- Community-oriented policing by problem solving
- Contemporary trend analysis for management
- Drug investigative techniques
- Electronic search and seizure
- Executive development
- Instructional techniques
- Major criminal investigative techniques
- Media communications
- Police explosive technicians
- Police labor relations
- Polygraph examiners
- Post–blast scene technicians
- Radiography
- Senior police administration
- Strategic intelligence analysis
- Systems approach to training design and delivery
- Telecommunication fraud investigative techniques

The Canadian approach is one of the leading approaches to police training in the world. Both the length of the training and the variety of topics included speak volumes about the seriousness of various experimentations. Probably the most impressive part of the training is the operational field training part, which exemplifies the complexity of police work and the constant need for additional in-service training. However, the Canadian police seem to suffer from a problem similar to that of U.S. forces. Despite the multitude of topics designed for in-service training, it appears that a follow-up mechanism that would enforce a systematic updated training does not yet exist.

GREAT BRITAIN

The British police force is composed of 43 forces in England and Wales in addition to the London Metropolitan Police. There are 27 county police forces, eight combined police areas, and six metropolitan police forces (Peak and Glensor, 1999; Rosenbaum, 1995).

Contrary to many assumptions, the police in Britain are not nationalized. Each regional force has its own chief constable (police chief), who is appointed for a term until retirement. The local police authority, consisting of citizens from the region, provides 49 percent of the funding for the police, and 51 percent comes through the British central government via the Home Office. A constable has law enforcement authority throughout England and Wales regardless of his or her employing force. Uniforms, police cars, and other police symbols are basically the same except for the center symbol of the force's badge or shield. Basic training curricula are consistent throughout England and Wales, with the home force providing instruction. Different forces have specialized training which is open to members of any force. To become a ranking officer of inspector or above, one must successfully complete a junior command course at the Police Staff College at Bramshill. Specialized management classes called *carousel classes* are also offered at Bramshill (Bayley, 1991). To become a constable or a line police officer, one must successfully complete the basic probationer training program (PTP). Information about the PTP is adapted from Police Training Council (1999). The British police training is under revision in 2001.

The PTP has been developed as a result of a review conducted in 1995–1996 and presented to the Police Training Council Executive Committee on October 17, 1996. The project brief was approved by the PTC on February 14, 1997. The aim of this training module was to provide probationers with the necessary skills, knowledge, attitude, and understanding to enable them to fulfill their future roles as police officers.

The probationary period in England normally lasts for two years, and during this time period probationers must demonstrate that they have become competent police officers. The probationary period consists of the following six stages.

Stage 1.   Stage 1 is held in force and is intended to be an introduction to policing and not merely an induction process. Although under very close supervision, probationers should be provided with the opportunity to observe work in police stations, familiarize themselves with basic equipment such as the personal radio, and ask questions about police work in general. In class they are introduced to local and national issues in preparation for stage 2 at a police training center (PTC). This stage last a minimum of two weeks.

Stage 2.   This stage of training takes place over 15 weeks at the PTC appropriate to the probationer's force area. Stage 2 provides a theoretical base for the entire program and includes simulations leading probationers toward "real" policing experience in subsequent force training. The emphasis is on student-centered participative skills-based application of law.

Stage 3.   Stage 3 consists of two weeks of in-force training which provides a force perspective and prepares officers for accompanied patrol. This may include locally determined skills training and reporting procedures.

Stage 4.   Stage 4 is held in-force and probationers work on a one-to-one basis with a trained tutor constable. They move on from stages 1, 2, and 3 to put into practice the skills and abilities they have developed. The tutor constable should deal with any complex situation that may arise, but when appropriate the probationer is encouraged to handle other matters under supervision. Stage 4, which provides a set of aims and objectives to give tutors some guidance, lasts for 10 weeks.

Stage 5.   Stage 5 consists of two weeks of in-force training which provides a local perspective and prepares officers for independent patrol. It will include all necessary force administration prior to undertaking formal independent patrol. The probationer should not attend this stage if decisions have been made to delay formal independent patrols for any reason.

Stage 6.   Stage 6 consists of a minimum of 20 days of formal training between weeks 32 and 104 of the probationary training. The actual format will vary from force to force. It is referred to as the *developmental stage*.

The British police, similar to some of its European counterparts, incorporates field training into basic academy training but should also be commended for additional innovation. The two weeks of introductory in-service training provides the desired plateau for familiarization with the idea of police. This introduction, made possible in the real-life environment of a given force, provides the recruit with a taste of " real police work" prior to basic academy training. A number of police departments in the United States experimented with this idea for a variety of reasons; however, it seems like a disappearing trend (Charleston, South Carolina, abandoned the pre-academy training). However, it should be considered seriously, especially by forces that do not have their own in-house academies.

JAPAN

The Japanese police force is divided into 47 prefectural police forces with approximately 220,0000 police officers (only 5400 women). The National Police Agency (NPA) is responsible for control and coordination of the prefectural police forces. While each prefecture determines its own policies and procedures, because the national police agency does not possess the legal standing to authorize nationally uniform of rules, the agency can suggest or guide the prefectures in the establishment of local policies and procedures.

The Metropolitan Police Force of Tokyo is headed by a superintendent general; the prefectural forces are administered by directors. There are several police stations within each prefecture that serve as the principal operational units of the police. Each station is subdivided into police (Bayley, 1991; Peak and Glensor, 1999). The National Police Agency delivers and supervises training to all prefecture employees (Reichel, 1999).

A police recruit can enter the service as either a police officer or an assistant inspector. For both positions the applicant is required to pass a national qualifying examination. Those recruited as police officers must have completed high school; approximately one-half of police officers are university graduates. Candidates for the rank of assistant inspector must have a college degree and must have passed an advanced civil service examination. As a whole, police officers are better educated than the Japanese population as a whole.

Candidates for police officer are recruited and trained at police schools in the prefectures. High school graduates spend one year at the school, and college graduates go through six months of training. Recruits are offered courses in the following areas:

- Law
- Police procedures
- Sociology
- Psychology
- History
- Literature
- Technical training

The second year of training is devoted to on-the-job training. New recruits are teamed with an experienced officer who has been selected to act as a tutor. The recruit then returns to the school for an additional six months to synthesize the theoretical training with the practical experiences of the job.

Applicants who have been recruited to the rank of assistant inspector spend six months in the training academy at the National Police Academy, where the program is designed to develop future police executives. In addition, there is an extensive system of special training designed to enhance officers' skills in particular aspects of police work. In-service courses, which prepare officers for promotion examinations are run by the regional police schools (Bayley, 1991).

In 1991, the National Police Agency embarked on a new policy to redesign the existing rank structure. The goal was to increase the number of officers above the rank of police sergeant to 40 percent of the total force, in recognition of the need for more officers with experience, in-depth knowledge, expertise, and specialization (Leishman, 1993). This recognition of the fact that expertise and specialization are the future of police organization in Japan should be commended, as should their extensive training program, which includes such innovative topics as history and literature. By inclusion of these two topics, the balance between old and new skills seems at least to be acknowledged.

HONG KONG

The material presented about the Hong Kong Police Department is based on an official publication of the Hong Kong Police Force (1999).

The Hong Kong National Police Force is commanded by the commissioner of police, who is responsible to the chief executive of the Hong Kong Special Administrative Region. The headquarters are divided into five departments: operations; crime and security; personnel and training; management services; and finance, administration, and planning. The strength of the force is 28,495 sworn officers and 6014 civilians.

The training wing, which is responsible for all matters related to training except training for internal security and the marine police, is composed of three divisions (pp. 11–12):

1. *In-Service Training Bureau:* responsible for providing developmental and vocational training
2. *Police Training Schools:* responsible for all recruit training and subsequent continuation and promotion training for all junior police officers
3. *Training Development Bureau:* responsible for the ultimate design, presentation, and review of all mainstream training materials, including the training of police officers who become police instructors

The In-Service Training Bureau is primarily responsible for command and management training and for a diverse vocational training, needed by officers of all ranks. Following is a list of courses offered by the bureau (pp. 53–54):

1. *Junior Command Course:* for inspectors and senior inspectors, a three-week preparatory course for the rank of chief inspector
2. *Intermediate Command Course:* a three-week course offered to newly promoted chief inspectors

3. *Senior Command Course:* a three-week course for recently promoted superintendents
4. *Police Driving School:* specialized course for all the ranks
5. *Weapons Training Division:* specialized course for all the ranks
6. *Auxiliary Training Division:* for the auxiliary police; length varies
7. *Detective Training School:* a specialized course divided into phases I and II

The Police Training School is responsible primarily for basic recruit training. It also provides some continuation and promotion training.

The Basic Training Division provides three types of basic training (pp. 54–55):

1. *Probationary Inspectors:* 36 weeks of basic training. The newly revised (1999) topics are:
   - Computer skills
   - Legal requirements
   - Procedural requirements
   - Human rights
   - Data privacy
2. *Recruit Police Constables:* 27 weeks of basic training. The newly introduced topics reflect the changing needs of the force and the community and include the following topics:
   - Law
   - Police and court procedures
   - Physical training
   - First aid
   - Lifesaving
   - Tactics
   - Foot drills
   - Firearms
   - Social studies
   - Resistance control

   Recruit police constables are also required to participate in a community interface program, to broaden their perspective and increase sensitivity to the needs of their community.
3. *Youth Preemployment Training Program:* government-sponsored program for school dropouts; lasts 2 weeks

Although the Hong Kong Police Force does not stand out with regard to the length of its training, it is certainly worth a second look based on its overall approach to training. Specifically, the specialized and advanced training offered on the premises of the police school provides an interesting model to learn from. The addition to the basic curriculum of such new topics as social studies and human rights validates the force's commitment to the community it serves.

SUMMARY

To summarize and complete this international review, each reader is encouraged to examine the training curricula of other countries, which provide a more comprehensive review of the various training concepts. They can be found on the Internet (*http://www.officeer.com*). Overall, a number of themes appear to be common to all the countries reviewed and unfortunately, do not seem to be taken under serious consideration in the United States, therefore providing some basic ideas about critical issues in police training. Following are some of the more visible themes:

1. *Duration of training:* much longer than that offered in the United States
2. *Subjects of instruction:* standardized for the entire police force
3. *Balance between "old" (practical) and "new" (conceptual) skills*
4. *FTO component incorporated into the academy training:* integration between theory and practice within the basic academy setting
5. *Advanced and specialized training offered by a centralized police training center*
6. *Expectations from graduates:* above average

The last theme appears to be the most critical one to be addressed by police academies in the United States, one that does not require additional allocation of resources but speaks volumes about recruitment and selection procedures. If we cannot recruit and select candidates who can perform at above average levels during basic academy training, something ought to be done to modify current recruitment and selection procedures.

REFERENCES

Bayley, D. H. (1991). *Forces of Order: Policing Modern Japan*, 2nd ed. Berkeley, CA: University of California Press.
Cerrah, Y. (2000). Personal communication. Turkish National Police Academy.
Dutch Ministry of Justice (1999). *Dutch Police Publication.*
Finnish Police (1999). *Police Training in Finland.* Official Publication of the National Police Academy.
Griffiths, C. T., Whitelaw, B., and Parent, R. B. (1999). *Canadian Police Work.* Scarborough, Ontario, Canada: International Thomson Publishing.
Haberfeld, M. (1999a). Field notes, Toronto, Ontario, Canada.
Haberfeld, M. (1999b). Field notes, Warsaw, Poland.
Haberfeld, M. (2000). Field notes, Warsaw, Poland.
Hong Kong Police Force (1999). Annual report.

Leishman, F. (1993). From pyramid to pencil: rank and rewards in Japan's police. *Police Studies, 16*, 33–36.

Michna, J. (1999). Polish police training. *Criminal Justice International*, October.

Murphy, P. J. (1996). *Law Enforcement News*, May.

Peak, K. J., and Glensor, R. W. (1999). *Community Policing and Problem Solving: Strategies and Practices*. Upper Saddle River, NJ: Prentice Hall.

Police College in Villingen-Schwenningen, Germany (1997). *Policing Overseas*.

Police Training Council (1999). *PTP Training Manual*. Bramshill College, England.

Reichel, P. L. (1999). *Comparative Criminal Justice Systems: A Topical Approach*. Upper Saddle River, NJ: Prentice Hall.

Rosenbaum, D. P. (1995). *The Challenge of Community Policing: Testing the Promises*. Thousand Oaks, CA: Sage Publications.

THE FUTURE
OF POLICE TRAINING
Small, Medium, or Large?

INTRODUCTION

The title of this chapter reflects some major considerations about the future of police training in the United States. The distinctions into small, medium, or large refer not only to department size (small, local with 50 sworn officers and below; medium, larger municipal, county, and state agencies with up to 1000 sworn officers; large, larger municipal, state, county, and federal organization that employ over 1000 officers) but also to the scale and range of the projected training.

The chapter begins with the introduction of three hypothetical models of training, followed by a theoretical discussion of research findings conducted in areas other than police training but clearly applicable to this topic, and culminates with a more pragmatic approach, listing some available sources of assistance and funding.

Entering the twenty-first century, depending on certain political trends, policy implementations, and hopefully, some impact of this book, police training can adapt one of the three models described below.

MODEL A: LARGE SCALE—UNIFORM PROFESSIONAL STANDARDS

Model A represents a large-scale approach—in a way, a revolutionary approach to existing training. It is predicated on implementation of the following recommendations:

1. Establish a federal POST commission that will be composed of all states' POST representatives, referred to as POPSC (Police Officers Professional Standards Commission), under the auspices of the Department of Justice.

2. Establish basic mandatory guidelines for all police agencies in the United States that will include clearly defined requirements for:

    (a) Topics to be included in the professional curriculum of a basic academy (Figure 15–1)

    (b) Length of each module (Figure 15–2)

    (c) Qualifications of the instructors

    (d) Recertification standards

    (e) In-service training

    (f) Developmental training

    (g) Advanced training

3. Establish police training centers or police schools, which in addition to state and local funding will be supplemented by resources from the federal budget.

4. Designate a professional association such as the ASLET, IDEALIST, or IACP to periodically assess and evaluate the professional training standards and recommend updates and changes.

5. Designate a professional association such as the American Society of Criminology or the Academy of Criminal Justice Sciences to form a permanent subcommittee to work with the organization designated in point 4 and the POPSC and to serve as a consulting body.

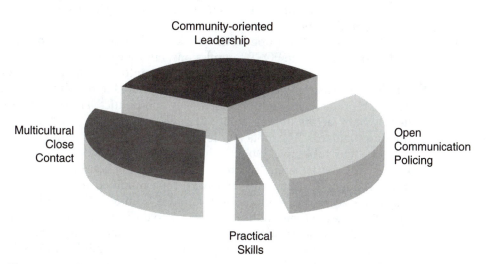

Figure 15–1   Proactive Training I

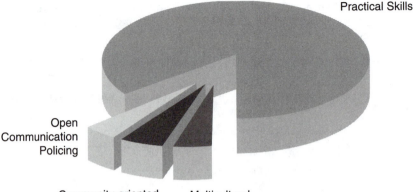

Figure 15-2   Proactive Training II

6. Establish mandatory first-line supervisor training program.
7. Establish a mandatory advanced training program for midlevel police executives, lieutenants, and captains.
8. Establish a mandatory advanced training program for police executives/upper management from the rank of major and above.

MODEL B: MEDIUM SCALE—REVISITING THE POST

Model B acknowledges the foreseeable criticism of model A—as an unconstitutional entity that takes away the right of each and every state to decide and addresses possible improvements in the presently existing situation with various POST commissions.

It is beyond the scope of this book to address all the problems associated with many of the state POST commissions, but they can be addressed in a generic manner. First, the problems can be divided into two categories: external and internal. External problems are in most cases related to two factors that have long been part of the history of policing: money and politics. The commissions are limited by both factors in their ability to devise all-inclusive, in-depth, professional training standards. The monetary limitations can also be divided into two factors: objective and subjective. Objectively speaking, there is only a certain amount of money that can be spent on police training by the employing agencies. Subjectively, though, this amount could have been modified if politics did not play a major role in resource allocation for police training. If our politicians devoted more time during their political campaigns to the issue of adequate police training and actually made some moves to allocate more money for police training, the commissions would probably know how to raise the minimum number of hours required of each training facility.

However, the politicians do not find the necessary urge to devote more time, energy, and resources to police training. Numerous reasons might be debatable, but this author has her own theory of why the issue of police training does not appear as critical to our politicians. The theory, referred to as the *conspiracy theory* (not to be confused with the conspiracy theory in the organized crime literature), presumes that nobody in general, but more specifically no politician, is seriously interested in raising policing into a true and reputable profession. The roots of this theory can be traced to the early history of policing (see Chapter 2), when the first designated police officer's task was actually to "clean up the mess" or coerce the berry gatherer to wake up and do his job. People in our society are divided into classes, and the ones who are in a way designated by us (through the choice of their occupation) to deal with our society's mess and dirty jobs do not, by default, belong to the upper class. If they did they would either have to hold professional status (attained through thorough education and training) or be compensated adequately, far above the average salary of police officer in today's world.

Both options pose some imminent danger. If police officers climb the ladder of social control and attain upper-class status through professional education, they might be less amenable to perform the tasks they are performing today without challenging the authority and rationality of the order imposed. Advanced education encourages critical thinking, critical thinking encourages challenging of the "obvious"—a police officer who challenges a given deployment order is not desirable material for a police force. Of course, some might argue that discipline has nothing to do with critical thinking, but this author disagrees entirely.

The internal problems facing the POST commissions are related to the staffing of any given committee. The author was once told by a colleague that if she wanted to improve the state of police training, she should volunteer her services to the local POST commission, as he does, instead of devoting her time to rewriting the rules and regulations of a small local police department. This personal example is brought here as an illustration of the narrowminded approach of an academic who views police training as an esoteric notion unrelated to what is happening inside police organizations. One cannot work only on the curriculum without assurance that police agencies in his or her jurisdiction are prepared for implementation of the curriculum. One can recommend mandatory roll-call training, but if this policy is not written into a department's rules and regulations, it has very little chance of success.

The problem with some of the decision makers who sit on POST committees is related to the fact that their understanding of training needs of police officers is either erroneous or, at best, extremely limited—they simply fail to see the "bigger picture." Some of the members are appointed based on their academic status, some based on the length of their career in law enforcement. This fundamentally incorrect approach to selection of POST members is related to the status of the academic profession on the one hand and the lack of status of the police profession on the other. It is assumed that a professor of criminal justice or police science will, by default, have an understanding of the police profession; the advanced degree will qualify him or her automatically as a valuable committee member. At the other end of the spectrum, senior police officers with 20 or 30 years of practical experience must also have an understanding of the police professions, and they, too, are valuable contributors. This approach, however, fails to recognize a simple fact—that to understand a profession and to have a "feel" for it is not a skill that is, by default, built into somebody's academic degree or years of service. One can hold a number of advanced degrees in the field of criminal justice and still have no practical understanding of police work. One can also be a police practitioner for 30 years and throughout those years not understand anything about the real nature of police work. This notion, by the way, holds true not only in the police profession but in any other profession as well. It takes passion, dedication, true devotion, and above all profound insight to be able to see the "bigger picture" beyond a mega label.

The recommendations for the newly revisited POST model include the following suggestions:

1. Form a recruitment and selection committee to identify possible candidates for POST commission membership.
2. Allocate sufficient monetary incentives to create full-time and well-paid positions for members of the POST commissions.
3. Mandate active participation of every police department in a given state by designating an officer whose part-time responsibility will include active cooperation with the POST commission.
4. Mandate the POST commissions to revise annually their minimum standards requirements for basic academy standards.
5. Mandate the POST commissions to devise and revise biannually minimum standards for first-line supervisor training.
6. Mandate the POST commissions to devise and revise biannually minimum standards for in-service specialized and advanced training.
7. Mandate the POST commissions to devise and revise periodically, but not less frequently than every three years, minimum standards for advanced training for midlevel police management.
8. Mandate the POST commissions to devise and revise periodically, but not less frequently than every five years, minimum standards for advanced training for upper police management.

MODEL C: SMALL SCALE—RELIVING THE PAST

Model C is probably the most realistic of the three models, even though the least desirable. The old adage "if it's not broken, don't fix it" seems to be the backbone of the training approach. Many police scholars and practitioners feel that tremendous advances have taken place over the last decades with regard to police training.

This author does not believe this to be an adequate portrayal of the state of current police training. This is not to say that many of our police officers are not exposed to much better and more sophisticated training than ever before. The problem, however, is not whether or not police officers today are exposed to virtual reality simulations during various use-of-force scenarios, but the themes mentioned throughout this book: critical issues of content, length, structure, audience, and so on.

The same tremendous advances have taken place in other areas of education as well. Medical students today are also trained in methods and tools unheard of a couple of decades ago. For medical students, however, these advances are an integral part of a well-oiled machine which represents the professional standards of the medical profession. Similar advances introduced to law enforcement personnel had, have and will continue to have only a very limited overall impact on police profession since they are introduced in a sporadic and haphazard manner. They do not represent another piece of a puzzle that will eventually contribute to the creation of a full picture as in the medical profession, for example. In policing, unfortunately, if we continue to approach training as segmented, unrelated, and "abbreviated" pieces, the advantages will not contribute to the creation of a full puzzle and will forever remain outside the puzzle's frame!

Nevertheless, beyond a philosophical theme, there must always be a more pragmatic alternative. Following are a number of recommendations to improve small-scale training efforts:

1. For small police departments without resources, mandatory subscriptions to periodicals such as *Law and Order* will provide training tips perfectly suited for roll-call training sessions, which can be delivered basically by any first-line supervisor or any line officer by taking turns.
2. Join ASLET and subscribe to its publications.
3. "Adopt" a professor, preferably an untenured one.
4. Provide the right environment.

The last recommendation constitutes a prelude to a much large problem, one that has not yet been addressed in this book and is highlighted in this chapter since its importance and amenability to change makes it a perfect candidate for a well-known exercise: "If you were to pick one thing out of all the concepts presented above, what would it be?" If the author had to pick just one thing, she would have concentrated on the area of creating the right environment. The reasons for picking this concept are related to the fact that with some minimal cost-effective effort, the entire approach to training can be changed in the most extreme manner.

Practically no serious research has been conducted in the area of creating the "right" physical and psychological environment that would optimize any given training effort. From personal observations the author noted that money, energy, and other resources have been wasted by introducing various training modules at the wrong time and in the wrong place.

Since the scarcity of available research does not allow for serious consideration of the topic, the following discussion is based on research conducted in the areas of physical work conditions in other environments, as the findings from other environments seem extremely applicable to training environments.

WORKING ENVIRONMENT

The physical working environment includes many factors and can have a major impact on any group of employees. The area has been researched extensively, and certain factors have been found to be more relevant than others. Sundstrom (1986) divides the physical environment into three different levels.

1. *Individual outcomes.* Important here is a person's satisfaction with environmental variables.
2. *Interpersonal outcomes.* The primary focus is on communication between employees, work performance, and interaction
3. *Organizational outcomes.* The primary focus is on the actual structure of the workplace and its effectiveness.

Level 1: Individual Outcomes

Level 1 consists of factors such as:

- Temperature
- Air
- Lighting
- Noise
- Music
- Color
- Equipment
- Space
- Rooms

Each of these factors can have a detrimental influence on work performance but can have a similar if not more extreme impact on the effectiveness of training delivered under the wrong circumstances.

Temperature and air can affect morale, efficiency, and physical well-being. Some people prefer cold weather, some like the heat. Some get depressed when they have no influence on these factors. Extensive research has been conducted on the temperature and air and the association with work performance. Studies on temperature suggest that 70°F (± 10°F) gives the best work performance outcome (Givoni and Rim, 1962; Weston, 1921).

Regardless of whether the training is conducted during the roll-call sessions, basic academies, or some other environments "customized" for training needs, attention to the factors presented above is crucial. Many training facilities are far from what can be described as the "right environments" for learning and absorbing information. Following is a brief overview of some of the factors identified by researchers and their potential impact on the effectiveness of training. With some minimal effort, by addressing some of the factors identified in this chapter, people responsible for training can make a huge difference.

Air-conditioning is not standard equipment in many training facilities, nor it is always found in briefing rooms. Officers exposed to excessive heat during a training session have diminished ability to learn and to absorb new concepts. Some attention devoted to this seemingly nonrelated training issue can contribute greatly to the effectiveness of the training module. If air-conditioning is too great an expense, training sessions have to be rescheduled and training offered when the temperature is expected to be closer to the ideal.

Smoking has recently been the most studied air pollutant. Jones and Reynolds (1978) found that nonsmokers tend to build up aggression and stress when smoking is present. Smoking is still extremely prevalent in law enforcement environments. Some departments attempt to create smoke-free environments, but many do not. Creating a smoke-free environment, at least for the duration of a training session, is certainly an attainable goal, especially considering the fact that it may deflect a nonsmoker's attention from concentrating on the uncomfortable feeling to focusing on what is being conveyed in the room.

Lighting can create another problem if it is inadequate in the work environment. Natural light (the sun) is difficult to influence for people working outside, but for inside workers, light bulbs can be adjusted. In a field study on lighting, Elton (1920) found that there is a correlation between lighting and production. Similar results were found in a study by Pressy (1921) in which students worked on different problems under different illumination conditions. A training session delivered in dark windowless rooms should be replaced with well-lighted environments whenever possible. It is quite amazing what difference a few additional light bulbs can make on a person's ability to absorb training material.

Noise can make people nervous and irritable if it becomes too loud, and for some workers it can become a physical hazard if the noise is continuously loud. Multiple studies have been conducted on the effects of noise in the work environment, and the research suggests that repeated loud, unpredictable, and uncontrollable noises can influence work performance both directly and indirectly (Glass and Singer, 1972). The author mentioned earlier that she was asked to deliver a sensitivity module to a group of graduating recruits. Among many distractions, related to the wrong environment, noise was probably the most detrimental for both students and the instructor. Especially when dealing with sensitive and/or complex topics, noise should be eliminated as much as possible. There are some training modules during which noise is pertinent to effective performance. For example, in a number of use-of-force scenarios, noise is certainly a valuable factor—but not during cultural diversity training.

Color can provide a pleasant work environment and may raise employee morale. According to Aaronson (1970) different colors represent associations with psychological states:

- *Red:* adventurous and sociable
- *Orange:* impulsive and quarrelsome
- *Yellow:* affectionate and impulsive
- *Blue:* affectionate and cautious
- *Violet:* gloomy and impulsive
- *Black:* gloomy and sullen
- *Gray:* cautious and gloomy
- *White:* shy and sociable

The author does not suggest that the walls of training academies and briefing rooms be painted red. However, most of those walls are painted in white or gray colors, creating a gloomy mood. Maybe a tiny splash of different color would contribute to a more cheerful environment that would inadvertently create a better learning atmosphere.

Music has been found to increase production slightly, and the claim is that employees become happier and more efficient when they listen to music. It is rather hard to envision training sessions for police officers accompanied by sounds of music. However, some training videos are accompanied by sounds that are either intentionally or by chance creating an atmosphere of intimidation. It is possible that when showing a video about police corruption, intimidating background music contributes to the desired learning effect. On the other hand, perhaps a less intimidating sound would have a more positive effect.

Overall, the key processes relating to a work environment that influence individual workers in level 1 are:

- Adaptation
- Arousal
- Overload
- Stress
- Mental fatigue
- Physiological fatigue
- Attitudes

Work satisfaction and performance are ultimately linked to these processes (Fischer, 1997). Effective learning is, quite clearly, related to the same concepts.

Level 2: Interpersonal Outcomes

The most important factor in level 2 is the physical layout of the work office. The work space includes the expression of self-identity, such as being able to have some control over one's own situation (personalization); additionally, work space also conveys status, which refers to the way that physical environment displays authority and influence (bigger office, rewards, etc).

The office layout's impact was also linked to small-group communication patterns. The research on communication has found that it is important to work in an environment, whether inside or outside, that promotes conversation, a face-to-face contact environment (Conrath, 1973). Privacy is related to confidentiality, which sometimes is necessary; however, too much privacy can result in isolation. Therefore, the work environment should try to address both the need for privacy, with, for example, conventional offices, and the need for conversation, such as open offices. A small-group model would be concerned with how offices can be arranged in a manner that would allow small groups of people to have both privacy and face-to-face contact if they are located in one office together.

The key processes that can be found in interpersonal relationships are identity and status, regulation of interaction, privacy, and choices in communication. Creating the right environments in these areas contributes to the creation of adequate communication and group cohesion (Fischer, 1997).

Training sessions, especially those delivered in-house in older police facilities, tend to represent exactly the same concepts of lack of status, privacy, confidentiality, and so on. The FBI National Academy is a prime example of a modern facility that, at least from a visual standpoint, encourages students to take seriously the time they spend in the lecture hall. A similar environment can be found in the International Law Enforcement Academy in Budapest, Hungary, where modular furniture in bright, well-lighted rooms practically "begs" for a serious, learning-conducive atmosphere. Closer to home, in local agencies, the East Brunswick, New Jersey, Police Department opened a new training facility in April 2000, which can be viewed by readers by accessing the EBPD Web site.

The facility is rather modest in size but extremely well designed and, probably unintentionally, addresses many of the factors discussed in this chapter. It is, of course, not always possible to modernize all the law enforcement facilities, but with some imagination and full awareness of the impact that a given facility might have, miracles can be performed.

Level 3: Organizational Outcomes

Level 3 deals with the structure and dynamics of the organization as a whole. The key process here is the congruence of organizational process and structure with the physical environment, an outcome of which influences the organizational effectiveness (Fisher, 1997). According to Fisher (1997), the physical aspects of the work environment are associated with three cognitive processes:

1. *Environmental stress:* ". . . the organism's psychological and physiological mobilization in response to an outside demand" (p. 54)
2. *Mental and physiological fatigue:* poor working performance due to fatigue
3. *Distraction:* shift in attention due to stressful events

Of course, none of the innovations discussed will have any real impact if the organization as a whole will not communicate to its employees where it stands in regard to professional development. If an organization does not feel that professional development constitutes an integral part of effective police work, people in charge of/or coordinating specialized, advanced, developmental, and roll-call training are treated as "outcasts" who carry this responsibility due to some "sins of the past" (as the author observed is the case in a number of police departments), and all the efforts put in place are wasted.

Prior to implementation of any of the concepts introduced in this chapter, all police organizations must make it clear that those in charge of training, whether directors of training in large in-house academies or designated line officers who are going to run roll-call training, be perceived as the ultimate professionals in the agency. In one department researched by the author, being a part of an FTO unit used to be perceived as an honor; unfortunately, things have changed. It is rather difficult to find an agency that communicates a message of respect and seriousness toward the personnel involved in training. Without elevating the status of training officers, there will be no effective police training, whether on a large or a small scale.

WORK PERFORMANCE

The second theoretical concept that needs to be addressed is derived from the area of organizational psychology and deals with work performance and its relation to motivational factors. This area of study also provides some insights into creation of the "right environment," which is so crucial to the positive outcome of any training effort.

Organizational psychology research has concluded that job performance depends on job satisfaction. For effective work to take place, a person needs to respond to the work environment in a favorable manner. However, satisfaction, in turn, depends on motivation. According to Cummings and Schwab (1973) there are three key principles involved in the motivational process. The degree to which are met provide insight into the way people perform:

1. Activation of a person's attention
2. Relationship between motivation and performance
3. Endurance of motivation

The researchers also define motivation as the process relating to the category of outcomes that a person wants to achieve or to avoid as well as to specific actions necessary to attain this.

Several different motivation theories have been developed over the years. Thierry (1998) identifies four categories of motivation theories:

1. *Content theories.* Focus is placed on what draws attention and what causes behavior to occur.
2. *Process theories.* Behavior is charged, channeled, and changed from a dynamic aspect.
3. *Reinforcement theories.* Motivation is developed from both positive (rewards) and negative (punishment) aspects, with the person as a passive organism.
4. *Cognitive theories.* The person is actively involved through his cognitive processing systems.

Work motivation can be described as the process that energizes and moves people, either intrinsically or extrinsically, toward a goal. At the end of the process lies job satisfaction. Satisfaction plays a role in every motivational theory, and it is important for successful work performance that satisfaction is achieved.

Lowenberg and Conrad (1998) discuss two sets of key correlates in job satisfaction: input and output correlates. The input correlates consist of individual characteristics and situational variables. The individual characteristics refer to how every person is unique and responds differently to situations. Examples are *personality*—differences occur within people and needs may change over time; *age*—job satisfaction is high in young and old age groups; and *gender and race*—minorities often show less job satisfaction, which may be a result of lack of opportunities for higher-level and more powerful jobs. The other input correlate, *situational variables*, refer to the relationship between the pay, supervision and organization, and job status (Lowenberg and Conrad, 1998).

Lowenberg and Conrad (1998) describe job performance as the main output correlate, and variables such as withdrawal, absenteeism, turnover, and theft as minor output correlates. Further, Schwab and Cummings (1973) identified three different theoretical points of view to explain the relationship:

1. Satisfaction leads to performance.
2. Satisfaction and performance are not directly related. Other variables are involved.
3. Performance leads to satisfaction, which in return can result in higher performance.

Another factor that influences work performance is shift work. Thierry and Jansen (1998, p. 109) concluded that shift work:

- Causes many subjective complaints concerning sleep, fatigue, and workload
- Leads to various nervous phenomena, such as headaches, depression, and trembling hands
- May cause problems with appetite and increase the chance of gastrointestinal complaints

- Probably contributes to the development of cardiovascular diseases
- Possibly has a detrimental effect on the working of the female reproductive system

The inclusion of an overview of a number of work performance motivational theories, discussed in this chapter, serves two purposes: first, to highlight the importance of effective training as a major correlate to effective job performance, and second, to offer an alternative way of thinking about the overall investment of resources into training of law enforcement officers.

The second point can be further expended by revisiting the post hoc training approach, introduced in Chapter 11. Regardless of the fact that some might argue with the accuracy of the depiction of specialized and developmental training as an alibi rather than a true professional development, nobody negates the fact that police work is complex in nature and has a significant impact on our society. Few will disagree with a notion that with all the efforts directed toward better selection and recruitment mechanisms, there is still a lot left to be desired.

The multitude of factors involved in the delivery of effective training modules is not an exclusive domain of law enforcement training. Any school teacher or college professor can identify the students who absorb the materials being taught better than others. The reasons for some students getting better grades than others is only secondary, as grades do not always reflect the true knowledge and understanding of a given topic. The author can point to any student in her class and identify the level of his or her knowledge of the topic being taught, regardless of the student's final grade.

Why is it, therefore, that some students get better grades than others, some students understand the topics better than others, and some simply fail to grasp anything? The answer, of course, is not simple. There are a number of factors that provide a full explanation, among them factors not discussed in this chapter, both internal and external to any person. However, part of the answer, perhaps even a significant part, is related to the work-performance and motivational theories.

A good class exercise to analyze each of the theories and factors discussed in this chapter and identify its correlation to effective police training. Following are some examples of the more obvious correlates:

- *Shift work* contributes not only to various forms of stress but also, inadvertently, to poor attention during roll-call training.
- *Poor pay and low status* can serve as motivators or demotivators: motivators if specialized training can be tied clearly to a monetary raise or to a higher status within the organization, demotivators if sending officers for developmental training (e.g., the RCPIs) is treated as another assignment that has to be fulfilled without any monetary or status benefit.

- *Input correlates*, if not taken into consideration when devising a specialized training module might cause overall dissatisfaction, even if expressed by only a few officers. Frequently, mixed groups exposed to training are not beneficial to the specific goal of a given training module (an example of which is discussed in Chapter 9).

- *Output correlates*, if not factored in, can invalidate the module, similar to the way that input correlates do.

The overall contribution of work performance and motivational theories to better understanding of police training cannot be underestimated. Training and education are probably among the most serious and complex tasks any person has to perform. It is, therefore, extremely unfortunate that law enforcement training and education are not addressed profoundly and seriously.

The final part of this chapter is devoted to some available resources which can be utilized by any law enforcement agency that views its training efforts as a matter of ultimate importance.

FUNDING FOR LAW ENFORCEMENT TRAINING PROGRAMS

The National Institute of Justice (NIJ) offers several awards in various areas of law enforcement. These areas can be viewed on the Internet (*http://www.ojp.usdoj.gov/nij*) and application forms are also available. NIJ also lists other Justice Department Organizations that offer funding to law enforcement.

Office of Juvenile Justice and Delinquency Prevention.   A complete list of grants and awards that are available through this organization can be found on the Internet (*http://www.ojjdp.ncjrs.org*). Any law enforcement agency can apply for the various funding opportunities.

Bureau of Justice Assistance.   This organization provides funding for several programs in law enforcement: Agricultural Crime Technology Information and Operations Network, Clandestine Drug Laboratory Enforcement Training, Clandestine Methaphetamine Laboratory Enforcement Training and Technical Assistance, Law Enforcement Services, State and Local Anti-Terrorism Training Program, and Tools for Tolerance. Information on areas of funding can be found on the internet (*http://www.ojp.usdoj.gov/BJA*).

Office of Justice Programs.   Various grants are offered to law enforcement agencies. Their programs can be downloaded from the World Wide Web (*http://www.ojp.usdoj.gov/fundopps.htm*).

Bureau of Justice Statistics.   The department offers funding in four broad areas (*http://www.Ojp.usdoj.gov/bjs/crs.htm*):

1. *National Criminal History Improvement Program (NCHIP).* Funding is provided to programs that improve the quality, timeliness, and immediate accessibility of criminal history and related records.

2. *National Sex Offender Registry Assistance Program.* This program is designed to assist in developing and enhancing sex offender registries, which can identify and contain current address information on convicted sex offenders who are released from supervision.

3. *National Incident-Based Reporting System Implementation Program (NIBRS).* Programs that have as an objective to improve the quality of crime statistics in the United States are funded.

4. *Criminal Records Policy Program.* This program is designed to support surveys, studies, conferences, and technical assistance on issues related to criminal justice records.

COPS MORE 2000.   The purpose of this program is to help police departments becoming more efficient by funding the support staff. The goal is to improve the training conditions for today's law enforcement officers (*http://www.cops.usdoj.gov/gen*).

Peer Support Training Institute (PSTI).   Training in peer support is offered to officers in smaller police departments (*http://www.peersupport.com/smallpeer.htm*). The purpose is train the program's participants to develop peer support programs in their respective departments. The curriculum includes the following topics:

- Techniques of relating and listening
- Consequences of stress
- Relationships
- Alcohol abuse
- Depression
- PTSD
- Grief and bereavement
- Stress management

In addition to the funding opportunities above, information regarding monetary assistance can also be found in the annual journal *Funding Law Enforcement Hotline* (*http://www.policecenter.com/bulletins/fp.shtml*). This publication gives a comprehensive listing of private and federal grants, funding news, project profiles, and step-by-step articles from law enforcement funding experts. Information includes deadlines, funds available, complete contact information, and important proposal requirements. A one-year subscription is $149.

SUMMARY

To achieve the status of professional is not an easy task. It takes a body of knowledge that can be absorbed, assessed, analyzed, and evaluated. Consistency in this body of knowledge is a paramount factor. Lawyers, accountants, social workers, and many other professionals are educated and trained in a consistent manner, regardless of the state or country of origin. In this chapter we have proposed a similar approach to the profession of policing by identifying the basic and mandatory standards and implementing them into every police training environment. The degree of customization remains open and rather flexible; however, the basic professional standards cannot be compromised.

One final note: Respect and high status for the police profession is not just a mirage—it is a real possibility. All it will take to make it happen is a lot of courage and vision, and since policing is a profession of courageous and visionary men and women, the task is already accomplished halfway!

REFERENCES

Aaronson, B. S. (1970). Some affective stereotypes of color. *International Journal of Symbology, 2,* 15–27.

Bureau of Justice Assistance. Retrieved September 5, 2000 from the World Wide Web: *http://www.ojp.usdoj.gov/BJA.*

Bureau of Justice Statistics. Retrieved September 5, 2000 from the World Wide Web: *http://www. Ojp.usdoj.gov/bjs/crs.htm.*

Conrath, C. W. (1973). Communication patterns, organizational structure, and man: some relationships. *Human Factors, 15,* 459–470.

COPS MORE 2000. Retrieved September 5, 2000 from the World Wide Web: *http://www.cops.usdoj.gov/gen.*

Cummings, L. L., and Schwab, D. P. (1973). *Performance in Organizations: Determinants and Appraisal.* Glenview, IL: Scott, Foresman.

Elton, P. M. (1920). A study of output in silk weaving during winter months. *Industrial Fatigue Research Board Report 9.* London: Her Majesty's Stationery Office.

Fischer, G. N. (1997). *Individuals and Environment: A Psychosocial Approach to Workspace.* New York: Walter de Gruyter.

Funding Law Enforcement Hotline. Retrieved September 5, 2000 from the World Wide Web: *http://www.policecenter.com/bulletins/fp.shtml.*

Givoni, B., and Rim, Y. (1962). Effects of the thermal environment and psychological factors upon subjects' responses and performance of mental work. *Ergonomics, 5,* 99–114.

Glass, D. C., and Singer, J. E. (1972). *Urban Stress: Experiments on Noise and Social Stressors.* San Diego, CA: Academic Press.

Jones, N. B., and Reynolds, V. (eds.) (1978). *Human Behavior and Adaptation.* London: Taylor and Francis. Distributed by Halsted Press.

Lowenberg G., and Conrad, K. (1998). *Current Perspectives in Industrial/Organizational Psychology.* Boston, MA: Allyn and Bacon.

National Institute of Justice. Retrieved September 5, 2000 from the World Wide Web: *http://www.ojp.usdoj.gov/nij.*

Office of Justice Programs. Retrieved September 5, 2000 from the World Wide Web: *http://www.ojp.usdoj.gov/fundopps.htm.*

Office of Juvenile Justice and Delinquency Prevention. Retrieved September 5, 2000 from the World Wide Web: *http://www.ojjdp.ncjrs.org.*

Peer Support Training Institute. Retrieved September 5, 2000 from the World Wide Web: *http://www.peersupport.com/smallpeer.htm.*

Pressy, S. L. (1921). The influence of color upon mental and motor efficiency. *American Journal of Psychology, 32,* 326–356.

Schwab, D. P., and Cummings, L. L. (1973). Theories of performance and satisfaction: a review. In W. E. Scott and L. L. Cummings (eds.), *Readings in Organizational Behavior and Human Performance.* Homewood, IL: Irwin, pp. 130–141.

Sundstrom, E. (1986). *Work Places: The Psychology of the Physical Environment in Offices and Factories.* New York: Cambridge University Press.

Thierry, H. (1998). Motivation and satisfaction. In P. J. D. Drenth, H. Thierry, and C. J. de Wolff (eds.), *Organizational Psychology.* East Sussex, England: Psychology Press, pp. 253–289.

Thierry, H., and Jansen, B. (1998). Work time and behaviour at work. In P. J. D. Drenth, H. Thierry, and C. J. de Wolff (eds.), *Work Psychology,* East Sussex, England: Psychology Press, pp. 89–119.

Weston, H. C. (1921). A study of efficiency in fine linen weaving. *Industrial Health (Fatigue) Research Board Report 20.* London: Her Majesty's Stationery Office.

INDEX